Progress in Mathematics
Volume 100

Michael E. Taylor

Pseudodifferential Operators and Nonlinear PDE

Birkhäuser
Boston • Basel • Berlin

Michael E. Taylor
Department of Mathematics
University of North Carolina
Chapel Hill, NC 27599

Library of Congress Cataloging-in-Publication Data

Taylor, Michael Eugene, 1946-
Pseudodifferential operators and nonlinear PDE / Michael E. Taylor.
p. cm. -- (Progress in mathematics ; v. 100)
ISBN 0-8176-3595-5 (alk. paper).
1. Pseudodifferential operators. 2. Differential equations, Nonlinear. I. Title. II. Series: Progress in mathematics (Boston, Mass.) : vol. 100.
QA329.7.T39 1991 91-18337
515'.7242--dc20 CIP

Printed on acid-free paper.

Second Printing 1993.

ISBN 0-8176-3595-5
ISBN 3-7643-3595-5

Typeset text prepared by the Author in AMSTEX.
Printed and bound by Edwards Brothers, Inc., Ann Arbor, Michigan
Printed in the USA.

9 8 7 6 5 4 3 2

CONTENTS

Research supported by NSF grant DMS 8801457.

Introduction

For the past 25 years the theory of pseudodifferential operators has played an important role in many exciting and deep investigations into linear PDE. Over the past decade, this tool has also begun to yield interesting results in nonlinear PDE. This paper is devoted to a summary and reconsideration of some uses of pseudodifferential operator techniques in nonlinear PDE.

We begin with a preliminary chapter reviewing pseudodifferential operators as a tool developed for the linear theory. This chapter sets down some of the fundamental results and defines a bit of notation. It is also intended to serve readers interested in nonlinear PDE but without prior experience with pseudodifferential operators, to acquaint them with the basics of the theory.

We then turn to an exposition of the further development which has applications to the nonlinear theory. One goal has been to build a bridge between two approaches which have been used in a number of papers written in the past decade, one being the theory of paradifferential operators, pioneered by Bony [Bo] and Meyer [M1], the other the study of pseudodifferential operators whose symbols have limited regularity.

The latter approach is a natural sucessor to classical devices of deriving estimates for linear PDE whose coefficients have limited regularity in order to obtain results on nonlinear PDE. Of the two approaches, it is initially the simpler. After making some general observations about symbols with limited smoothness and their associated operators in §1.1, we illustrate this in §1.2, using very little machinery to derive some regularity results for solutions to nonlinear elliptic PDE. The results there assume a priori that the solutions have a fair amount of regularity. To obtain better results, harder work is required. One useful tool is the symbol decomposition studied in §1.3. The idea is to write a nonsmooth symbol $p(x,\xi)$ as $p^{\#}(x,\xi) + p^{b}(x,\xi)$ in such a way that an operator algebra is available for the associated operator $p^{\#}(x,D)$ while $p^{b}(x,D)$ is regarded as a remainder term to be estimated.

Chapter 2 establishes needed estimates on operators with non-smooth symbols. The material here incorporates ideas of Bourdaud [BG], Kumanogo and Nagase [KN], Marschall [Ma], and Meyer [M1]. Having these estimates, we return to elliptic PDE in §2.2, obtaining full strength Schauder estimates, though of course not the special estimates (for scalar second order elliptic PDE) of de Giorgi et al. and of Krylov and Safanov.

The symbol smoothing developed in §1.3 and applied in Chapter 2 provides a transition to the theory of paradifferential operators, which we expose in Chapter 3. In this chapter we take a further look at nonlinear elliptic PDE. Also, in §3.6 we use the paradifferential operator calculus to prove some commutator estimates, including important estimates of Coifman and Meyer [CM] and of Kato and Ponce [KP].

Chapter 4 exploits commutator estimates of Coifman and Meyer established in §3.6 to derive a sharp operator calculus for C^1 symbols, including some classical results of Calderon. We also compare this material with a C^1 paradifferential calculus.

In subsequent chapters we treat various basic topics in nonlinear PDE. Chapter 5 deals with nonlinear hyperbolic systems. We endeavor to obtain the sharpest results on regular solutions to symmetric and symmetrizable systems, though generalized solutions involving shock waves and such are not considered.

In Chapter 6 we establish a variant of Bony's propagation of singularities theorem. We mention one point; in showing that for a solution $u \in H^{m+\sigma}$ to a nonlinear PDE $F(x, D^m u) = f$, microlocal regularity of order $m + \sigma - 1 + s$ propagates for $s < r$, we require $u \in C^{m+r}$ rather than $u \in H^{\frac{1}{2}n+m+r}$. This implication is slightly more precise than the usual statement, and also highlights the mechanism giving rise to the higher regularity. It will be clear that this material could have been put right after Chapter 3, but since wave propagation is the basic phenomenon inducing one to be interested in propagation of singularities, it seems natural to put the material here.

Chapters 7 and 8 treat nonlinear parabolic equations and elliptic boundary problems, respectively. The latter topic extends the interior analysis done in Chapters 2 and 3. In both of these chapters we discuss some existence theorems which follow by using the DeGiorgi-Nash-Moser theory in concert with the results proved here. These arguments are well known but are included in order to help place in perspective what is done in these chapters. Also in Chapter 7 we derive some results on semilinear parabolic equations and illustrate these results by discussing how they apply to work on harmonic mappings. In Chapter 9 we discuss Nirenberg's refinement of the Schauder estimates for elliptic equations, and describe how this plays a role in the work of [CNS] on the Monge-Ampere equation.

In various applications to PDE we find different advantages in the diverse techniques developed in Chapters 1-4. For example, for interior elliptic regularity, symbol smoothing is very useful, but in the quasilinear case both $C^r S^m_{1,0}$-calculus as developed in Chapter 2 and paradifferential operator calculus as developed in Chapter 3 seem equally effective, while in the completely nonlinear case the latter tool seems to work better. The paradifferential approach is used on elliptic boundary problems in Chapter 8. For nonlinear hyperbolic equations, the sharpest results seem to be produced by a combination of the Kato-Ponce inequality and some generalizations, and $C^1 S^m_{cl}$-calculus, developed in Chapter 4. It will be noted that both these tools have roots in paradifferential operator calculus.

At the end are four appendices. The first collects some facts about various function spaces, particularly Sobolev spaces, Hölder spaces, and Zygmund spaces, and also Morrey spaces and BMO. The second ties together some known results and discusses a few new results on norm estimates of the form

$$\|u\|_X \le C\|u\|_Y \Big[1 + \log\big(\|u\|_Z/\|u\|_Y\big)\Big]^a,$$

in borderline cases when the inclusion $Y \subset X$ barely fails. This particularly arises in a number of important cases where either X or Y is L^∞. The third appendix gives a proof of the DeGiorgi-Nash-Moser estimates, largely following works of Moser and Morrey. Appendix D presents a proof of some paraproduct estimates of [CM] needed for some of the results of §3.5-§3.6, such as the commutator estimates used to develop the $C^1S^m_{cl}$-calculus in Chapter 4.

We also include a notational index, since a rather large number of function spaces and operator spaces arise naturally during the course of the investigations described here.

Chapter 0: Pseudodifferential operators and linear PDE

In this preliminary chapter we give an outline of the theory of pseudodifferential operators as it has been developed to treat problems in linear PDE, and which will provide a basis for further developments to be discussed in the following chapters. Many results will be proved in detail, but some proofs are only sketched, with references to more details in the literature. We define pseudodifferential operators with symbols in Hörmander's classes $S^m_{\rho,\delta}$, derive some useful properties of their Schwartz kernels, discuss their algebraic properties, then show how they can be used to establish regularity of solutions to elliptic PDE with smooth coefficients. We proceed to a discussion of mapping properties on L^2 and on the Sobolev spaces H^s, then discuss Gårding's inequality, and some of its refinements, known as sharp Gårding inequalities. In §0.8 we apply some of the previous material to establish existence of solutions to hyperbolic equations. We introduce the notion of wave front set in §0.10 and discuss microlocal regularity of solutions to elliptic equations. We also discuss how solution operators to a class of hyperbolic equations propagate wave front sets. In §0.11 we discuss L^p estimates, particularly some fundamental results of Calderon and Zygmund, and applications to Littlewood-Paley Theory, which will be an important technical tool for basic estimates established in Chapter 2. We end this introduction with a brief discussion of pseudodifferential operators on manifolds.

§0.1. The Fourier integral representation and symbol classes

The Fourier inversion formula is

$$f(x) = \int \hat{f}(\xi)\, e^{ix\cdot\xi}\, d\xi \tag{0.1.1}$$

where $\hat{f}(\xi) = (2\pi)^{-n} \int f(x) e^{-ix\cdot\xi} dx$ is the Fourier transform of a function on $\mathbb{R}^n$. If one differentiates (0.1.1), one obtains

$$D^\alpha f(x) = \int \xi^\alpha \hat{f}(\xi) e^{ix\cdot\xi} d\xi, \tag{0.1.2}$$

where $D^\alpha = D_1^{\alpha_1} \cdots D_n^{\alpha_n}$, $D_j = (1/i)\partial/\partial x_j$. Hence, if

$$p(x, D) = \sum_{|\alpha|\le k} a_\alpha(x) D^\alpha$$

is a differential operator, we have

$$p(x, D) f(x) = \int p(x,\xi) \hat{f}(\xi) e^{ix\cdot\xi} d\xi \tag{0.1.3}$$

where

$$p(x,\xi)=\sum_{|\alpha|\le k} a_\alpha(x)\xi^\alpha.$$

One uses the Fourier integral representation (0.1.3) to define pseudodifferential operators, taking the function $p(x,\xi)$ to belong to one of a number of different classes of symbols. In this chapter we consider the following symbol classes, first defined by Hörmander.

Assuming $\rho,\delta\in[0,1]$, $m\in\mathbb{R}$, we define $S^m_{\rho,\delta}$ to consist of C^∞ functions $p(x,\xi)$ satisfying

$$|D_x^\beta D_\xi^\alpha p(x,\xi)|\le C_{\alpha\beta}\langle\xi\rangle^{m-\rho|\alpha|+\delta|\beta|} \tag{0.1.4}$$

for all α, β, where $\langle\xi\rangle=(1+|\xi|^2)^{\frac{1}{2}}$. In such a case we say the associated operator defined by (0.1.3) belongs to $OPS^m_{\rho,\delta}$.

If there are smooth $p_{m-j}(x,\xi)$, homogeneous in ξ of degree $m-j$ for $|\xi|\ge 1$, i.e., $p_{m-j}(x,r\xi)=r^{m-j}p_{m-j}(x,\xi)$ for $r,|\xi|\ge 1$, and if

$$p(x,\xi)\sim\sum_{j\ge 0}p_{m-j}(x,\xi) \tag{0.1.5}$$

in the sense that

$$p(x,\xi)-\sum_{j=0}^{N}p_{m-j}(x,\xi)\in S^{m-N}_{1,0} \tag{0.1.6}$$

for all N, then we say $p(x,\xi)\in S^m_{cl}$, or just $p(x,\xi)\in S^m$.

It is easy to see that if $p(x,\xi)\in S^m_{\rho,\delta}$, and $\rho,\delta\in[0,1]$, then $p(x,D):\mathcal{S}(\mathbb{R}^n)\to C^\infty(\mathbb{R}^n)$. In fact, multiplying (0.1.3) by x^α, writing $x^\alpha e^{ix\cdot\xi}=(-D_\xi)^\alpha e^{ix\cdot\xi}$ and integrating by parts yields

$$p(x,D):\mathcal{S}(\mathbb{R}^n)\longrightarrow\mathcal{S}(\mathbb{R}^n). \tag{0.1.7}$$

Further mapping properties will be described below, but for now we make note of the following.

LEMMA 0.1.A. *If $\delta<1$, then*

$$p(x,D):\mathcal{S}'(\mathbb{R}^n)\longrightarrow\mathcal{S}'(\mathbb{R}^n). \tag{0.1.8}$$

PROOF: Given $u\in\mathcal{S}'$, $v\in\mathcal{S}$, we have formally

$$\langle v,p(x,D)u\rangle=\langle p_v,\hat{u}\rangle \tag{0.1.9}$$

where

$$p_v(\xi) = (2\pi)^{-n} \int v(x) p(x,\xi) e^{ix\cdot\xi} dx.$$

Now integration by parts gives

$$\xi^\alpha p_v(\xi) = (2\pi)^{-n} \int D_x^\alpha \big(v(x)p(x,\xi)\big) e^{ix\cdot\xi} dx,$$

so

$$|p_v(\xi)| \le C_\alpha \langle\xi\rangle^{m+\delta|\alpha|-|\alpha|}.$$

Thus if $\delta < 1$ we have rapid decrease of $p_v(\xi)$. Similarly we get rapid decrease of derivatives of $p_v(\xi)$, so it belongs to $\mathcal{S}$. Thus the right side of (0.1.9) is well defined.

As the case $\delta = 1$ will be very important in later chapters, the failure of (0.1.8) in this case will have definite consequences.

A useful alternative representation for a pseudodifferential operator is obtained via a synthesis of the family of unitary operators

$$e^{iq\cdot X}\ e^{ip\cdot D}\ u(x) = e^{iq\cdot x}\ u(x+p). \tag{0.1.10}$$

Given $a(x,\xi) \in \mathcal{S}(\mathbb{R}^n \times \mathbb{R}^n)$, we have

$$\begin{aligned}
&\int \hat{a}(q,p) e^{iq\cdot X} e^{ip\cdot D} u(x)\, dq dp \\
&= (2\pi)^{-2n} \int a(y,\xi) e^{-iy\cdot q} e^{-i\xi\cdot p} e^{ix\cdot q} u(x+p)\, dy d\xi dq dp \\
&= (2\pi)^{-n} \int a(x,\xi) e^{-i\xi\cdot p} u(x+p)\, d\xi dp \\
&= (2\pi)^{-n} \int a(x,\xi) e^{ix\cdot\xi} \hat{u}(\xi)\, d\xi.
\end{aligned} \tag{0.1.11}$$

In other words,

$$a(x,D)u = \int \hat{a}(q,p) e^{iq\cdot X} e^{ip\cdot D} u(x)\, dq dp. \tag{0.1.12}$$

This can be compared to the Weyl calculus, defined by

$$\begin{aligned}
a(X,D)u &= \int \hat{a}(q,p) e^{i(q\cdot X + p\cdot D)} u(x)\, dq dp \\
&= (2\pi)^{-n} \int a\Big(\frac{1}{2}(x+y),\xi\Big) e^{i(x-y)\cdot\xi} u(y)\, dy d\xi,
\end{aligned} \tag{0.1.13}$$

which has been extensively studied (see [H1]), but will not be used here.

§0.2. Schwartz kernels of pseudodifferential operators

To an operator $p(x, D) \in OPS^m_{\rho,\delta}$ defined by (0.1.3) corresponds a Schwartz kernel $K \in \mathcal{D}'(\mathbb{R}^n \times \mathbb{R}^n)$, satisfying

$$
\begin{aligned}
\langle u(x)v(y), K\rangle &= \iint u(x)p(x,\xi)\hat{v}(\xi)e^{ix\cdot\xi}d\xi dx \\
&= (2\pi)^{-n}\iiint u(x)p(x,\xi)e^{i(x-y)\cdot\xi}v(y)dy\, d\xi\, dx.
\end{aligned} \tag{0.2.1}
$$

Thus, K is given as an 'oscillatory integral'

$$
K = (2\pi)^{-n}\int p(x,\xi)e^{i(x-y)\cdot\xi}d\xi. \tag{0.2.2}
$$

We have the following basic result.

PROPOSITION 0.2.A. *If $\rho > 0$, then K is C^∞ off the diagonal in $\mathbb{R}^n \times \mathbb{R}^n$.*

PROOF: For given $\alpha \geq 0$,

$$
(x-y)^\alpha K = \int e^{i(x-y)\cdot\xi}\; D^\alpha_\xi p(x,\xi)\; d\xi. \tag{0.2.3}
$$

This integral is clearly absolutely convergent for $|\alpha|$ so large that $m-\rho|\alpha| < -n$. Similarly it is seen that applying j derivatives to (0.2.3) yields an absolutely convergent integral provided $m + j - \rho|\alpha| < -n$, so in that case $(x-y)^\alpha K \in C^j(\mathbb{R}^n \times \mathbb{R}^n)$. This gives the proof.

Generally, if T has the mapping properties

$$
T : C_0^\infty(\mathbb{R}^n) \longrightarrow C^\infty(\mathbb{R}^n), \quad T : \mathcal{E}'(\mathbb{R}^n) \longrightarrow \mathcal{D}'(\mathbb{R}^n)
$$

and its Schwartz kernel K is C^∞ off the diagonal, it follows easily that

$$
\text{sing supp } Tu \subset \text{ sing supp } u \text{ for } u \in \mathcal{E}'(\mathbb{R}^n).
$$

This is called the *pseudolocal property*. By (0.1.7)-(0.1.8) it holds for $T \in OPS^m_{\rho,\delta}$ if $\rho > 0$ and $\delta < 1$.

We remark that the proof of Proposition 0.2.A leads to the estimate

$$
|D^\beta_{x,y}K| \leq C|x-y|^{-k} \tag{0.2.4}
$$

where $k \geq 0$ is any integer strictly greater than $(1/\rho)(m + n + |\beta|)$. In fact, this estimate is rather crude. It is of interest to record a more precise estimate which holds when $p(x,\xi) \in S^m_{1,\delta}$.

PROPOSITION 0.2.B. *If* $p(x,\xi) \in S^m_{1,\delta}$, *then the Schwartz kernel* K *of* $p(x,D)$ *satisfies estimates*

$$|D^\beta_{x,y} K| \leq C|x-y|^{-n-m-|\beta|} \tag{0.2.5}$$

provided $m + |\beta| > -n$.

The result is easily reduced to the case $p(x,\xi) = p(\xi)$, satisfying $|D^\alpha p(\xi)| \leq C_\alpha \langle\xi\rangle^{m-|\alpha|}$, for which $p(D)$ has Schwartz kernel $K = \hat{p}(y-x)$. It suffices to prove (0.2.5) for such a case, for $\beta = 0$ and $m > -n$. We make use of the following simple but important characterization of such symbols.

LEMMA 0.2.C. *Given* $p(\xi) \in C^\infty(\mathbb{R}^n)$, *it belongs to* $S^m_{1,0}$ *if and only if*

$$p_r(\xi) = r^{-m} p(r\xi) \text{ is bounded in } C^\infty(1 \leq |\xi| \leq 2) \text{ for } r \in [1,\infty). \tag{0.2.6}$$

Given this, we can write $p(\xi) = p_0(\xi) + \int_0^\infty q_\tau(e^{-\tau}\xi)d\tau$ with $q_0(\xi) \in C_0^\infty(\mathbb{R}^n)$ and $e^{-m\tau}q_\tau(\xi)$ bounded in the Schwartz space $\mathcal{S}(\mathbb{R}^n)$, for $\tau \in [0,\infty)$. Hence $e^{-m\tau}\hat{q}_\tau(z)$ is bounded in $\mathcal{S}(\mathbb{R}^n)$. In particular, we have $e^{-m\tau}|\hat{q}_\tau(z)| \leq C_N\langle z\rangle^{-N}$, so

$$\begin{aligned} |\hat{p}(z)| &\leq |\hat{p}_0(z)| + C_N \int_0^\infty e^{(n+m)\tau}\left(1 + |e^\tau z|\right)^{-N} d\tau \\ &\leq C + C_N|z|^{-n-m} \int_{\log|z|}^\infty e^{(n+m)\tau}(1+e^\tau)^{-N} d\tau, \end{aligned} \tag{0.2.7}$$

which implies (0.2.5). We also see that in the case $m + |\beta| = -n$, we obtain a result upon replacing the right side of (0.2.5) by $C \log|x-y|^{-1}$, (provided $|x-y| < \frac{1}{2}$).

§0.3. Adjoints and products

Given $p(x,\xi) \in S^m_{\rho,\delta}$, we obtain readily from the definition that

$$p(x,D)^* v = (2\pi)^{-n} \int p(y,\xi)^* e^{i(x-y)\cdot\xi} v(y)dyd\xi. \tag{0.3.1}$$

This is not quite in the form (0.1.3) as the amplitude $p(y,\xi)^*$ is not a function of (x,ξ). Before continuing the analysis of (0.3.1), we are motivated to look at a general class of operators

$$Au(x) = (2\pi)^{-n} \int a(x,y,\xi) e^{i(x-y)\cdot\xi} u(y)dyd\xi. \tag{0.3.2}$$

We assume

$$|D_y^\gamma D_x^\beta D_\xi^\alpha a(x,y,\xi)| \le C_{\alpha\beta\gamma}\langle\xi\rangle^{m-\rho|\alpha|+\delta_1|\beta|+\delta_2|\gamma|}, \tag{0.3.3}$$

and then say $a(x,y,\xi) \in S^m_{\rho,\delta_1,\delta_2}$. A brief calculation transforms (0.3.2) to

$$(2\pi)^{-n}\int q(x,\xi)e^{i(x-y)\cdot\xi}u(y)\,dy\,d\xi \tag{0.3.4}$$

with

$$\begin{aligned} q(x,\xi) &= (2\pi)^{-n}\int a(x,y,\eta)e^{i(x-y)\cdot(\eta-\xi)}\,dy\,d\eta \\ &= e^{iD_\xi\cdot D_y}a(x,y,\xi)|_{y=x}. \end{aligned} \tag{0.3.5}$$

Note that a formal expansion $e^{iD_\xi\cdot D_y} = I + iD_\xi\cdot D_y - \frac{1}{2}(D_\xi\cdot D_y)^2 + \cdots$ gives

$$q(x,\xi) \sim \sum_{\alpha\ge 0}\frac{i^{|\alpha|}}{\alpha!}D_\xi^\alpha D_y^\alpha a(x,y,\xi)|_{y=x}. \tag{0.3.6}$$

If $a(x,y,\xi)\in S^m_{\rho,\delta_1,\delta_2}$ with $0\le\delta_2<\rho\le 1$, then the general term in (0.3.6) belongs to $S^{m-(\rho-\delta)|\alpha|}_{\rho,\delta}$, $\delta=\min(\delta_1,\delta_2)$, so the sum on the right is formally asymptotic. This suggests the following result:

PROPOSITION 0.3.A. *If $a(x,y,\xi)\in S^m_{\rho,\delta_1,\delta_2}$ with $0\le\delta_2<\rho\le 1$, then the operator (0.3.2) belongs to $OPS^m_{\rho,\delta}$, with $\delta=\max(\delta_1,\delta_2)$. In fact $A = q(x,D)$ where $q(x,\xi)$ has the asymptotic expansion (0.3.6).*

To prove this proposition, one can first show that the Schwartz kernel $K(x,y) = (2\pi)^{-n}\int a(x,y,\xi)e^{i(x-y)\cdot\xi}d\xi$ satisfies the same estimates as established in Proposition 0.2.A, and hence, altering A only by an operator in $OPS^{-\infty}$, we can assume $a(x,y,\xi)$ is supported on $|x-y|\le 1$. Let

$$\hat{b}(x,\eta,\xi) = (2\pi)^{-n}\int a(x,x+y,\xi)e^{-iy\cdot\eta}\,dy, \tag{0.3.7}$$

so

$$p(x,\xi) = \int \hat{b}(x,\eta,\xi+\eta)\,d\eta. \tag{0.3.8}$$

The hypotheses on $a(x,y,\xi)$ imply

$$|D_x^\beta D_\eta^\alpha \hat{b}(x,\eta,\xi)| \le C_{\nu\alpha\beta}\langle\xi\rangle^{m+\delta|\beta|+\delta_2\nu-\rho|\alpha|}\langle\eta\rangle^{-\nu} \tag{0.3.9}$$

where $\delta = \max\,(\delta_1, \delta_2)$. Since $\delta_2 < 1$, it follows that $p(x,\xi)$ and any of its derivatives can be bounded by some power of $\langle\xi\rangle$.

Now a power series expansion of $\hat{b}(x,\eta,\xi+\eta)$ in the last argument about ξ gives

(0.3.10)
$$\Big|\hat{b}(x,\eta,\xi+\eta) - \sum_{|\alpha|<N} \frac{1}{\alpha!}(iD_\xi)^\alpha \hat{b}(x,\eta,\xi)\eta^\alpha\Big| \le C_\nu |\eta|^N \langle\eta\rangle^{-\nu} \sup_{0\le t\le 1} \langle\xi + t\eta\rangle^{m+\delta_2\nu-\rho N}.$$

With $\nu = N$ we get a bound

$$C\langle\xi\rangle^{m-(\rho-\delta_2)N} \text{ if } |\eta| \le \frac{1}{2}|\xi|, \tag{0.3.11}$$

and if ν is large we get a bound by any power of $\langle\eta\rangle^{-1}$ for $|\xi| < 2|\eta|$. Hence

(0.3.12)
$$\Big|p(x,\xi) - \sum_{|\alpha|<N} \frac{1}{\alpha!}(iD_\xi)^\alpha D_y^\alpha a(x,x+y,\xi)|_{y=0}\Big| \le C\langle\xi\rangle^{m+n-(\rho-\delta_2)N},$$

from which the proposition follows.

If we apply Proposition 0.3.A to (0.3.1), we obtain:

Proposition 0.3.B. *If $p(x,D) \in OPS^m_{\rho,\delta}$, $0 \le \delta < \rho \le 1$, then*

$$p(x,D)^* = p^*(x,D) \in OPS^m_{\rho,\delta} \tag{0.3.13}$$

with

$$p^*(x,\xi) \sim \sum_{\alpha\ge 0} \frac{i^{|\alpha|}}{\alpha!} D_\xi^\alpha D_x^\alpha p(x,\xi)^*. \tag{0.3.14}$$

The result for products of pseudodifferential operators is the following.

Proposition 0.3.C. *Given $p_j(x,\xi) \in OPS^{m_j}_{\rho_j,\delta_j}$, suppose*

$$0 \le \delta_2 < \rho \le 1 \text{ with } \rho = \min(\rho_1,\rho_2). \tag{0.3.15}$$

Then

$$p_1(x,D)p_2(x,D) = q(x,D) \in OPS^{m_1+m_2}_{\rho,\delta} \tag{0.3.16}$$

with $\delta = \max(\delta_1,\delta_2)$, and

$$q(x,\xi) \sim \sum_{\alpha\ge 0} \frac{i^{|\alpha|}}{\alpha!} D_\xi^\alpha p_1(x,\xi)\; D_x^\alpha p_2(x,\xi). \tag{0.3.17}$$

This can be proved by writing

$$p_1(x,D)p_2(x,D)u = p_1(x,D)p_2^*(x,D)^*u = Au \tag{0.3.18}$$

for A as in (0.3.2) with

$$a(x,y,\xi) = p_1(x,\xi)p_2^*(y,\xi)^* \tag{0.3.19}$$

and then applying Proposition 0.3.A and 0.3.B. Alternatively, one can compute directly that $p_1(x,D)p_2(x,D) = q(x,D)$ with

$$\begin{aligned} q(x,\xi) &= (2\pi)^{-n} \int p_1(x,\eta)p_2(y,\xi)e^{i(x-y)\cdot(\eta-\xi)}\,d\eta dy \\ &= e^{iD_\eta\cdot D_y}p_1(x,\eta)p_2(y,\xi)\big|_{y=x,\eta=\xi}, \end{aligned} \tag{0.3.20}$$

and then apply an analysis such as used to prove Proposition 0.3.A. Carrying out this latter approach has the minor advantage that the hypothesis (0.3.15) can be weakened to $0 \le \delta_2 < \rho_1 \le 1$, which is quite natural since the right side of (0.3.17) is formally asymptotic under such a hypothesis.

§0.4. Elliptic operators and parametrices

We say $p(x,D) \in OPS^m_{\rho,\delta}$ is elliptic if, for some $r < \infty$,

$$|p(x,\xi)| \ge C\langle\xi\rangle^m \text{ for } |\xi| \ge r. \tag{0.4.1}$$

Thus, if $\psi(\xi) \in C^\infty(\mathbb{R}^n)$ is equal to 0 for $|\xi| \le r$, 1 for $|\xi| \ge 2r$, it follows easily from the chain rule that

$$\psi(\xi)p(x,\xi)^{-1} = q_0(x,\xi) \in S^{-m}_{\rho,\delta}. \tag{0.4.2}$$

As long as $0 \le \delta < \rho \le 1$, we can apply Proposition 0.3.C to obtain

$$\begin{aligned} q_0(x,D)p(x,D) &= I + r_0(x,D) \\ p(x,D)q_0(x,D) &= I + \tilde{r}_0(x,D) \end{aligned} \tag{0.4.3}$$

with

$$r_0(x,\xi),\ \tilde{r}_0(x,\xi) \in S^{-(\rho-\delta)}_{\rho,\delta}. \tag{0.4.4}$$

Using the formal expansion

$$I - r_0(x,D) + r_0(x,D)^2 - \cdots \sim I + s(x,D) \in OPS^0_{\rho,\delta} \tag{0.4.5}$$

and setting $q(x,D) = (I = s(x,D))q_0(x,D) \in OPS^{-m}_{\rho,\delta}$, we have

(0.4.6) $$q(x,D)p(x,D) = I + r(x,D), \quad r(x,\xi) \in S^{-\infty}.$$

Similarly we obtain $\tilde{q}(x,D) \in OPS^{-m}_{\rho,\delta}$ satisfying

(0.4.7) $$p(x,D)\tilde{q}(x,D) = I + \tilde{r}(x,D), \quad \tilde{r}(x,\xi) \in S^{-\infty}.$$

But evaluating

(0.4.8) $$\big(q(x,D)p(x,D)\big)\tilde{q}(x,D) = q(x,D)\big(p(x,D)\tilde{q}(x,D)\big)$$

yields $q(x,D) = \tilde{q}(x,D) \bmod OPS^{-\infty}$, so in fact

(0.4.9) $$\begin{aligned} q(x,D)p(x,D) &= I \bmod OPS^{-\infty} \\ p(x,D)q(x,D) &= I \bmod OPS^{-\infty}. \end{aligned}$$

We say $q(x,D)$ is a two-sided parametrix for $p(x,D)$.

The parametrix can establish local regularity of a solution to

(0.4.10) $$p(x,D)u = f.$$

Suppose $u, f \in \mathcal{S}'(\mathbb{R}^n)$, and $p(x,D) \in OPS^m_{\rho,\delta}$ is elliptic, with $0 \le \delta < \rho \le 1$. Constructing $q(x,D) \in OPS^{-m}_{\rho,\delta}$ as in (0.4.6), we have

(0.4.11) $$u = q(x,D)f - r(x,D)u.$$

Now a simple analysis parallel to (0.1.7) implies that

(0.4.12) $$R \in OPS^{-\infty} \Longrightarrow R : \mathcal{E}' \longrightarrow \mathcal{S}.$$

By duality, since taking adjoints preserves $OPS^{-\infty}$,

(0.4.13) $$R \in OPS^{-\infty} \Longrightarrow R : \mathcal{S}' \longrightarrow C^\infty.$$

Thus (0.4.11) implies

(0.4.14) $$u = q(x,D)f \bmod C^\infty.$$

Applying the pseudolocal property to (0.4.10) and (0.4.14), we have the following elliptic regularity result.

PROPOSITION 0.4.A. *If $p(x,D) \in OPS^m_{\rho,\delta}$ is elliptic and $0 \le \delta < \rho \le 1$, then, for any $u \in \mathcal{S}'(\mathbb{R}^n)$,*

(0.4.15) $$\textit{sing supp } p(x,D)u = \textit{ sing supp } u.$$

More refined elliptic regularity involves keeping track of Sobolev space regularity. As we have the parametrix, this will follow simply from mapping properties of pseudodifferential operators, to be established in subsequent sections.

§0.5. L^2 estimates

Here we want to obtain L^2 estimates for pseudodifferential operators. The following simple basic estimate will get us started.

PROPOSITION 0.5.A. *Let (X, μ) be a measure space. Suppose $k(x, y)$ is measurable on $X \times X$ and*

$$\int_X |k(x,y)| d\mu(x) \leq C_1, \quad \int_X |k(x,y)| d\mu(y) \leq C_2, \tag{0.5.1}$$

for all y and x, respectively. Then

$$Tu(x) = \int k(x,y) u(y) d\mu(y) \tag{0.5.2}$$

satisfies

$$\|Tu\|_{L^p} \leq C_1^{\frac{1}{p}} \ C_2^{\frac{1}{q}} \ \|u\|_{L^p} \tag{0.5.3}$$

for $p \in [1, \infty]$, with $\frac{1}{p} + \frac{1}{q} = 1$.

PROOF: For $p \in (1, \infty)$, estimate

$$\left| \iint k(x,y) u(y) v(x) d\mu(y) d\mu(x) \right| \tag{0.5.4}$$

via $|uv| \leq |u(x)|^p/p + |v(y)|^q/q$. Then (0.5.4) is dominated by $(C_1/p)\|u\|_{L^p}^p + (C_2/q)\|v\|_{L^q}^q$. Replacing u, v by $tu, t^{-1}v$ and minimizing the resulting bound over $t \in (0, \infty)$, we dominate (0.5.4) by $C_1^{1/p} C_2^{1/q} \|u\|_{L^p} \|v\|_{L^q}$, thus proving (0.5.3). The exceptional cases $p = 1$ and ∞ are easily treated.

To apply this when $X = \mathbb{R}^n$ and $k = K$ is the Schwartz kernel of $p(x, D) \in OPS^m_{\rho,\delta}$, note from the proof of Proposition 0.2.A that

$$|K(x,y)| \leq C_N |x-y|^{-N} \text{ for } |x-y| \geq 1 \tag{0.5.5}$$

as long as $\rho > 0$, while

$$|K(x,y)| \leq C|x-y|^{-(n-1)} \text{ for } |x-y| \leq 1 \tag{0.5.6}$$

as long as $m < -n + \rho(n-1)$. (Recall this last estimate is actually rather crude.) Hence we have the following preliminary result.

LEMMA 0.5.B. *If $p(x,D) \in OPS^m_{\rho,\delta}$, $\rho > 0$, and $m < -n + \rho(n-1)$, then*

$$p(x,D) : L^p(\mathbb{R}^n) \longrightarrow L^p(\mathbb{R}^n), \quad 1 \le p \le \infty. \tag{0.5.7}$$

If $p(x,D) \in OPS^m_{1,\delta}$, then (0.5.7) holds for $m < 0$.

The last observation follows from the improvement of (0.5.6) given in (0.2.5).

Our main goal in this section is to prove the following.

THEOREM 0.5.C. *If $p(x,D) \in OPS^0_{\rho,\delta}$ and $0 \le \delta < \rho \le 1$, then*

$$p(x,D) : L^2(\mathbb{R}^n) \longrightarrow L^2(\mathbb{R}^n). \tag{0.5.8}$$

The proof we give, following [H4], begins with:

LEMMA 0.5.D. *If $p(x,D) \in OPS^{-a}_{\rho,\delta}$, $0 \le \delta < \rho \le 1$, and $a > 0$, then (0.5.8) holds.*

PROOF: Since $\|Pu\|^2_{L^2} = (P^*Pu, u)$, it suffices to prove that some power of $p(x,D)^*p(x,D) = Q$ is bounded on L^2. But $Q^k \in OPS^{-2ka}_{\rho,\delta}$, so for k large enough this follows from Lemma 0.5.B.

To proceed with the proof of Theorem 0.5.C, taking

$$q(x,D) = p(x,D)^*p(x,D) \in OPS^0_{\rho,\delta},$$

suppose $|q(x,\xi)| \le M - b,\ b > 0$, so

$$M - \operatorname{Re} q(x,\xi) \ge b > 0. \tag{0.5.9}$$

In the matrix case, take $\operatorname{Re} q(x,\xi) = \frac{1}{2}(q(x,\xi) + q(x,\xi)^*)$. It follows that

$$A(x,\xi) = \big(M - \operatorname{Re} q(x,\xi)\big)^{\frac{1}{2}} \in S^0_{\rho,\delta} \tag{0.5.10}$$

and

$$\begin{aligned} A(x,D)^*A(x,D) &= M - q(x,D) + r(x,D), \\ & r(x,D) \in OPS^{-(\rho-\delta)}_{\rho,\delta}. \end{aligned} \tag{0.5.11}$$

Applying Lemma 0.5.D to $r(x,D)$, we have

$$\begin{aligned} M\|u\|^2_{L^2} - \|p(x,D)u\|^2_{L^2} &= \|A(x,D)u\|^2_{L^2} - (r(x,D)u,u) \\ &\ge -C\|u\|^2_{L^2} \end{aligned} \tag{0.5.12}$$

or

$$\|p(x,D)u\|^2 \le (M+C)\|u\|_{L^2}^2, \tag{0.5.13}$$

finishing the proof.

From these L^2-estimates easily follow L^2-Sobolev space estimates. The Sobolev space $H^s(\mathbb{R}^n)$ is often defined as

$$H^s(\mathbb{R}^n) = \{u \in \mathcal{S}'(\mathbb{R}^n) : \langle\xi\rangle^s \hat{u}(\xi) \in L^2(\mathbb{R}^n)\}. \tag{0.5.14}$$

Equivalently, with

$$\Lambda^s u = \int \langle\xi\rangle^s \hat{u}(\xi) e^{ix\cdot\xi}\,d\xi; \quad \Lambda^s \in OPS^s, \tag{0.5.15}$$

we have

$$H^s(\mathbb{R}^n) = \Lambda^{-s} L^2(\mathbb{R}^n). \tag{0.5.16}$$

The operator calculus easily gives:

PROPOSITION 0.5.E. *If* $p(x,D) \in OPS^m_{\rho,\delta}$, $0 \le \delta < \rho \le 1$, $m, s \in \mathbb{R}$, *then*

$$p(x,D) : H^s(\mathbb{R}^n) \longrightarrow H^{s-m}(\mathbb{R}^m). \tag{0.5.17}$$

Note that, in view of the boundedness of operators with symbols $\xi^\alpha\langle\xi\rangle^{-k}$, $|\alpha| \le k \in \mathbb{Z}^+$, and $\langle\xi\rangle^{2k}\big[\sum_{|\alpha|\le k} |\xi^\alpha|^2\big]^{-1}$, we easily see that, given $u \in L^2(\mathbb{R}^n)$, then $u \in H^k(\mathbb{R}^n)$ if and only if $D^\alpha u \in L^2(\mathbb{R}^n)$ for $|\alpha| \le k$, so the definition (0.5.16) is consistent with other common notions of H^k when $s = k \in \mathbb{Z}^+$.

Given Proposition 0.5.E, one easily obtains Sobolev regularity of solutions to the elliptic equations studied in §0.4.

Calderon and Vaillancourt sharpened Theorem 0.5.C, showing that

$$p(x,\xi) \in S^0_{\rho,\rho},\ 0 \le \rho < 1 \Longrightarrow p(x,D) : L^2(\mathbb{R}^n) \longrightarrow L^2(\mathbb{R}^n). \tag{0.5.18}$$

This result, particularly for $\rho = \frac{1}{2}$, has played an important role in linear PDE, especially in the study of subelliptic operators, but it will not be used in this paper.

§0.6. Gårding's inequality

In this section we establish a fundamental estimate, first obtained by Gårding in the case of differential operators.

THEOREM 0.6.A. *If* $p(x,D) \in OPS^m_{\rho,\delta}$, $0 \le \delta < \rho \le 1$, *and*

$$\text{Re } p(x,\xi) \ge C|\xi|^m \text{ for } |\xi| \text{ large,} \tag{0.6.1}$$

then, for any $s \in \mathbb{R}$, *there are* C_0, C_1 *such that*

$$\text{Re } (p(x,D)u,u) \ge C_0\|u\|^2_{H^{m/2}} - C_1\|u\|^2_{H^s}. \tag{0.6.2}$$

PROOF: Replacing $p(x,D)$ by $\Lambda^{-m/2}p(x,D)\Lambda^{-m/2}$, we can suppose without loss of generality that $m = 0$. Then, as in the proof of Theorem 0.5.C, take

$$A(x,\xi) = \left(\text{Re } p(x,\xi) - \frac{1}{2}C\right)^{\frac{1}{2}} \in S^0_{\rho,\delta}, \tag{0.6.3}$$

so

$$\begin{aligned} A(x,D)^*A(x,D) = {} & \text{Re } p(x,D) - \frac{1}{2}C + r(x,D), \\ & r(x,D) \in OPS^{-(\rho-\delta)}_{\rho,\delta}. \end{aligned} \tag{0.6.4}$$

This gives

$$\begin{aligned} \text{Re } (p(x,D)u,u) &= \|A(x,D)u\|^2_{L^2} + \frac{1}{2}C\|u\|^2_{L^2} + (r(x,D)u,u) \\ &\ge \frac{1}{2}C\|u\|^2_{L^2} - C_1\|u\|^2_{H^s} \end{aligned} \tag{0.6.5}$$

with $s = -(\rho-\delta)/2$, so (0.6.2) holds in this case. If $s < -(\rho-\delta)/2 = s_0$, use the simple estimate

$$\|u\|^2_{H^s} \le \varepsilon\|u\|^2_{L^2} + C(\varepsilon)\|u\|^2_{H^{s_0}} \tag{0.6.6}$$

to obtain the desired result in this case.

§0.7. The sharp Gårding inequality

In this section we will sketch a proof of the following sharp Gårding inequality, first proved by Hörmander for scalar operators and then by Lax and Nirenberg for matrix valued operators.

PROPOSITION 0.7.A. *If* $p(x,\xi) \in S^1_{1,0}$ *and* $p(x,\xi) \ge 0$, *then*

$$\text{Re } (p(x,D)u,u) \ge -C\|u\|^2_{L^2}. \tag{0.7.1}$$

We begin with the following characterization of $S^m_{1,0}$, a variant of (0.2.6). Cover $\mathbb{R}^{2n}$ with 'rectangles' R_j, centered at points (x_j, ξ_j), such that $|x - x_j| \leq \frac{1}{2}$ and $|\xi - \xi_j| \leq \frac{1}{2}M_j$ defines R_j; here $M_j = \max(|\xi_{j1}|, \ldots, |\xi_{jn}|)$, if this max is ≥ 1, $M_j = 2$ otherwise; $\xi_j = (\xi_{j1}, \ldots, \xi_{jn})$. Let Ψ_j be the natural affine map from the unit 'cube' Q_0 in $\mathbb{R}^{2n}$, defined by $|x| \leq \frac{1}{2}$, $|\xi| \leq \frac{1}{2}$, onto R_j. Then $p(x,\xi) \in C^\infty(\mathbb{R}^{2n})$ defines a sequence of functions $p \circ \Psi_j \in C^\infty(Q_0)$, and $p(x,\xi)$ belongs to $S^m_{1,0}$ if and only if $\{M_j^{-m} p \circ \Psi_j\}$ is bounded in $C^\infty(Q_0)$.

One can pick a cover R_j of $\mathbb{R}^{2n}$ and a subordinate partition of unity $\psi_j(x,\xi) \geq 0$, bounded in $S^0_{1,0}$. Let

$$q_j(x,\xi) = (\psi_j p) \circ \Psi_j. \tag{0.7.2}$$

Then $p(x,\xi) \in S^1_{1,0}$ implies $M_j^{-1} q_j$ bounded in $C_0^\infty(Q_0)$.

Now one can construct 'by hand' an operator $A = a_1(x, D)$, such that $a_1(x,\xi) \in \mathcal{S}(\mathbb{R}^{2n})$, $\int a_1(x,\xi)dxd\xi = 1$, and $(Au, u) \geq 0$ for all u. One can take A to be given by (0.3.2) with

$$a(x,y,\xi) = C_0 e^{-|x|^2} e^{-|\xi|^2} e^{-|y|^2} \tag{0.7.3}$$

and verify that this works, for some $C_0 > 0$. Then set

$$a_t(x,\xi) = t^{n/2} a(t^{1/2}x, t^{1/2}\xi) \tag{0.7.4}$$

and define $\tilde{p}_j \in \mathcal{S}(\mathbb{R}^{2n})$ by

$$\tilde{p}_j \circ \Psi_j = q_j * a_{M_j}. \tag{0.7.5}$$

Thus

$$\tilde{p}_j = (\psi_j p) * b_j \tag{0.7.6}$$

where b_j is obtained from $a_1(x,\xi)$ via a linear symplectic transformation; $b_j(x, D)$ is unitarily conjugate to $a_1(x, D)$, and hence $b_j(x, D) \geq 0$. Since $(\psi_j p)(x,\xi) \geq 0$, it follows from (0.7.6) that

$$\tilde{p}_j(x, D) = \int (\psi_j p)(y,\eta) b_j(x - y, D - \eta) dy d\eta \geq 0.$$

Now let

$$\tilde{p}(x,\xi) = \sum_j \tilde{p}_j(x,\xi). \tag{0.7.7}$$

It is clear that $\tilde{p}(x, D) \geq 0$. It is also not hard to show that

$$p(x,\xi) - \tilde{p}(x,\xi) \in S^0_{1,0}. \tag{0.7.8}$$

This gives (0.7.1).

Note that, multiplying $p(x, D)$ on both sides by $\Lambda^{m/2}$, you can restate the sharp Gårding inequality in the apparently more general form: if $p(x,\xi) \in S^m_{1,0}$ is ≥ 0, then

$$\text{Re}\,(p(x,D)u,u) \geq -C\|u\|^2_{H^{(m-1)/2}}. \tag{0.7.9}$$

There is also the following variant for symbols of type (ρ, δ) :

PROPOSITION 0.7.B. *Given* $0 \leq \delta < \rho \leq 1$, *then* $p(x,\xi) \in S^m_{\rho,\delta}$, $p(x,\xi) \geq 0$ *implies* (0.7.1) *provided* $m = \rho - \delta$.

The proof is parallel to that sketched above. One covers $\mathbb{R}^{2n}$ by rectangles R_j, defined by $|x - x_j| \leq \frac{1}{2}M_j^{-\delta}$, $|\xi - \xi_j| \leq \frac{1}{2}M_j^{\rho}$.

For scalar operators, there is the following tremendous strengthening, due to Fefferman and Phong [FP]:

THEOREM 0.7.C. *If* $p(x,\xi) \in S^m_{\rho,\delta}$, *is* ≥ 0 *and scalar, then* (0.7.1) *holds provided* $m = 2(\rho - \delta)$.

A variant of the sharp Gårding inequality, for symbols with limited smoothness, will be established in Chapter 2 and applied in Chapter 6. The approach in Chapter 2 will be to write $p(x,\xi) = p^{\#}(x,\xi)+p^{b}(x,\xi)$ and apply Proposition 0.7.B to $p^{\#}(x,\xi)$, while bounding the norm of $p^{b}(x,\xi)$. Actually, in order to state the sharpest available result, we will apply Theorem 0.7.C to $p^{\#}(x,\xi)$, but for the application in Chapter 6 the weaker result following from Proposition 0.7.B suffices.

§0.8. Hyperbolic evolution equations

In this section we examine first order systems of the form

$$\partial u/\partial t = L(t,x,D_x)u + g(t,x), \quad u(0) = f. \tag{0.8.1}$$

We assume $L(t,x,\xi) \in S^1_{1,0}$, with smooth dependence on t, so

$$|D_t^j D_x^\beta D_\xi^\alpha L(t,x,\xi)| \leq C_{j\alpha\beta}\langle\xi\rangle^{1-|\alpha|}. \tag{0.8.2}$$

Here $L(t,x,\xi)$ is a $K \times K$ matrix-valued function, and we make the hypothesis of symmetric hyperbolicity:

$$L(t,x,\xi)^* + L(t,x,\xi) \in S^0_{1,0}. \tag{0.8.3}$$

We suppose $f \in H^s(\mathbb{R}^n)$, $s \in \mathbb{R}$, $g \in C(\mathbb{R}, H^s(\mathbb{R}^n))$.

Our strategy will be to obtain a solution to (0.8.1) as a limit of solutions u_ε to

$$\partial u_\varepsilon/\partial t = J_\varepsilon L J_\varepsilon u_\varepsilon + g, \quad u_\varepsilon(0) = f, \tag{0.8.4}$$

where

$$J_\varepsilon = \varphi(\varepsilon D_x) \tag{0.8.5}$$

for some $\varphi(\xi) \in \mathcal{S}(\mathbb{R}^n)$, $\varphi(0) = 1$. The family of operators J_ε is called a Friedrichs mollifier. Note that, for any $\varepsilon > 0$, $J_\varepsilon \in OPS^{-\infty}$, while, for $\varepsilon \in (0,1]$, J_ε is bounded in $OPS^0_{1,0}$.

For any $\varepsilon > 0$, $J_\varepsilon L J_\varepsilon$ is a bounded linear operator on each H^s, and solvability of (0.8.4) is elementary. Our next task is to obtain estimates on u_ε, independent of $\varepsilon \in (0,1]$. Use the norm $||u||_{H^s} = ||\Lambda^s u||_{L^2}$. We derive an estimate for

$$(d/dt)||\Lambda^s u_\varepsilon(t)||^2_{L^2} = 2(\Lambda^s J_\varepsilon L J_\varepsilon u_\varepsilon) + 2(\Lambda^s g, \Lambda^s u_\varepsilon). \tag{0.8.6}$$

Write the first two terms on the right as

$$2(L\Lambda^s J_\varepsilon u_\varepsilon, \Lambda^s J_\varepsilon u_\varepsilon) + 2([\Lambda^s, L] J_\varepsilon u_\varepsilon, \Lambda^s J_\varepsilon u_\varepsilon). \tag{0.8.7}$$

By (0.8.3), $L + L^* = B(t,x,D) \in OPS^0_{1,0}$, so the first term in (0.8.7) is equal to

$$(B(t,x,D)\Lambda^s J_\varepsilon u_\varepsilon, \Lambda^s J_\varepsilon u_\varepsilon) \le C||J_\varepsilon u_\varepsilon||^2_{H^s}. \tag{0.8.8}$$

Meanwhile, $[\Lambda^s, L] \in OPS^s_{1,0}$, so the second term in (0.8.7) is also bounded by the right side of (0.8.8). Applying Cauchy's inequality to $2(\Lambda^s g, \Lambda^s u_\varepsilon)$, we obtain

$$(d/dt)||\Lambda^s u_\varepsilon(t)||^2_{L^2} \le C||\Lambda^s u_\varepsilon(t)||^2_{L^2} + C||g(t)||^2_{H^s}. \tag{0.8.9}$$

Thus Gronwall's inequality yields an estimate

$$||u_\varepsilon(t)||^2_{H^s} \le C(t)\left[||f||^2_{H^s} + ||g||^2_{C([0,t],H^s)}\right], \tag{0.8.10}$$

independent of $\varepsilon \in (0,1]$. We are now prepared to establish the following existence result.

PROPOSITION 0.8.A. *If* (0.8.1) *is symmetric hyperbolic and*

$$f \in H^s(\mathbb{R}^n), \ g \in C(\mathbb{R}, H^s(\mathbb{R}^n)), \ s \in \mathbb{R},$$

then there is a solution u *to* (0.8.1), *satisfying*

$$u \in L^\infty_{\text{loc}}(\mathbb{R}, H^s(\mathbb{R}^n)) \cap \ Lip(\mathbb{R}, H^{s-1}(\mathbb{R}^n)). \tag{0.8.11}$$

PROOF: Take $I = [-T, T]$. The bounded family

$$u_\varepsilon \in C(I, H^s) \cap C^1(I, H^{s-1})$$

will have a weak limit point u satisfying (0.8.11), and it is easy to verify that such u solves (0.8.1). As for the bound on $[-T, 0]$, this follows from invariance of the class of hyperbolic equations under time reversal.

Analogous energy estimates can establish uniqueness of such a solution u and rates of convergence of $u_\varepsilon \to u$ as $\varepsilon \to 0$. Also (0.8.11) can be improved to

$$u \in C(\mathbb{R}, H^s(\mathbb{R}^n)) \cap C^1(\mathbb{R}, H^{s-1}(\mathbb{R}^n)). \tag{0.8.12}$$

As details of such arguments, applied to nonlinear problems, can be found in Chapter 5, we will skip them here.

There are other notions of hyperbolicity. In particular, (0.8.1) is said to be symmetrizable hyperbolic if there is a $K \times K$ matrix valued $S(t, x, \xi) \in S^0_{1,0}$ which is positive definite, and such that $S(t, x, \xi)L(t, x, \xi) = \tilde{L}(t, x, \xi)$ satisfies (0.8.3). It can be shown that (0.8.1) is symmetrizable whenever it is strictly hyperbolic, i.e., if $L \in S^1_{cl}$ and $L_1(t, x, \xi)$ has, for each (t, x, ξ), $\xi \neq 0$, K distinct purely imaginary eigenvalues. Proposition 0.8.A extends to the case of symmetrizable hyperbolic systems. First order systems of this nature arise from higher order strictly hyperbolic PDE. Such arguments as justify these statements can also be found, applied to nonlinear problems, in Chapter 5.

§0.9. Egorov's Theorem

We want to examine the behavior of operators obtained by conjugating a pseudodifferential operator $P_0 \in OPS^m_{1,0}$ by the solution operator to a scalar hyperbolic equation of the form

$$\partial u / \partial t = iA(t, x, D_x)u, \tag{0.9.1}$$

where we assume $A = A_1 + A_0$ with

$$A_1(t, x, \xi) \in S^1_{cl} \text{ real}, \quad A_0(t, x, \xi) \in S^0_{cl}. \tag{0.9.2}$$

We suppose $A_1(t,x,\xi)$ is homogeneous in ξ, for $|\xi| \geq 1$. Denote by $S(t,s)$ the solution operator to (0.9.1), taking $u(s)$ to $u(t)$. This is a bounded operator on each Sobolev space H^σ, with inverse $S(s,t)$. Set

$$P(t) = S(t,0)P_0S(0,t). \tag{0.9.3}$$

We aim to prove the following result of Egorov.

THEOREM 0.9.A. *If $P_0 = p_0(x,D) \in OPS^m_{1,0}$, then for each t, $P(t) \in OPS^m_{1,0}$, modulo a smoothing operator. The principal symbol of $P(t)$ (mod $S^{m-1}_{1,0}$) at a point (x_0,ξ_0) is equal to $p_0(y_0,\eta_0)$, where (y_0,η_0) is obtained from (x_0,ξ_0) by following the flow $\mathcal{C}(t)$ generated by the (time dependent) Hamiltonian vector field*

$$H_{A_1(t,x,\xi)} = \sum_{j=1}^n \Big(\frac{\partial A_1}{\partial \xi_j}\frac{\partial}{\partial x_j} - \frac{\partial A_1}{\partial x_j}\frac{\partial}{\partial \xi_j}\Big). \tag{0.9.4}$$

To start the proof, differentiating (0.9.3) with respect to t yields

$$P'(t) = i[A(t,x,D),P(t)], \quad P(0) = P_0. \tag{0.9.5}$$

We will construct an approximate solution $Q(t)$ to (0.9.5) and then show that $Q(t) - P(t)$ is a smoothing operator.

So we are looking for $Q(t) = q(t,x,D) \in OPS^m_{1,0}$, solving

$$Q'(t) = i[A(t,x,D),Q(t)] + R(t), \quad Q(0) = P_0, \tag{0.9.6}$$

where $R(t)$ is a smooth family of operators in $OPS^{-\infty}$. We do this by constructing the symbol $q(t,x,\xi)$ in the form

$$q(t,x,\xi) \sim q_0(t,x,\xi) + q_1(t,x,\xi) + \cdots. \tag{0.9.7}$$

Now the symbol of $i[A,Q(t)]$ is of the form

$$H_{A_1}q + \{A_0,q\} + i\sum_{|\alpha|\geq 2} \frac{i^{|\alpha|}}{\alpha!}\Big(A^{(\alpha)}q_{(\alpha)} - q^{(\alpha)}A_{(\alpha)}\Big) \tag{0.9.8}$$

where $A^{(\alpha)} = D^\alpha_\xi A$, $A_{(\alpha)} = D^\alpha_x A$, etc. Since we want the difference between this and $\partial q/\partial t$ to have order $-\infty$, this suggests defining $q_0(t,x,\xi)$ by

$$\Big(\frac{\partial}{\partial t} - H_{A_1}\Big)q_0(t,x,\xi) = 0, \quad q_0(0,x,\xi) = p_0(x,\xi). \tag{0.9.9}$$

Thus $q_0(t, x_0, \xi_0) = p_0(y_0, \eta_0)$, as in the statement of the Theorem; therefore $q_0(t, x, \xi) \in S^m_{1,0}$. The equation (0.9.9) is called a *transport equation.* Recursively we obtain transport equations

$$\left(\frac{\partial}{\partial t} - H_{A_1}\right) q_j(t, x, \xi) = b_j(t, x, \xi), \quad q_j(0, x, \xi) = 0, \tag{0.9.10}$$

for $j \geq 1$, with solutions in $S^{m-j}_{1,0}$, leading to a solution to (0.9.6).

Finally we show $P(t) - Q(t)$ is a smoothing operator. Equivalently, we show that, for any $f \in H^\sigma(\mathbb{R}^n)$,

$$v(t) - w(t) = S(t,0)P_0 f - Q(t)S(t,0)f \in H^\infty(\mathbb{R}^n), \tag{0.9.11}$$

where $H^\infty(\mathbb{R}^n) = \cap_s H^s(\mathbb{R}^n)$. Note that

$$\partial v/\partial t = iA(t, x, D)v, \quad v(0) = P_0 f \tag{0.9.12}$$

while use of (0.9.6) gives

$$\partial w/\partial t = iA(t, x, D)w + g, \quad w(0) = P_0 f \tag{0.9.13}$$

where

$$g = R(t)S(t,0)w \in C^\infty(\mathbb{R}, H^\infty(\mathbb{R}^n)). \tag{0.9.14}$$

Hence

$$(\partial/\partial t)(v - w) = iA(t, x, D)(v - w) - g, \quad v(0) - w(0) = 0. \tag{0.9.15}$$

Thus energy estimates for hyperbolic equations yield $v(t) - w(t) \in H^\infty$ for any $f \in H^\sigma(\mathbb{R}^n)$, completing the proof.

A check of the proof shows that

$$P_0 \in OPS^m_{cl} \Longrightarrow P(t) \in OPS^m_{cl} \tag{0.9.16}$$

Also the proof readily extends to yield the following:

PROPOSITION 0.9.B. *With $A(t, x, D)$ as before,*

$$P_0 \in OPS^m_{\rho,\delta} \Longrightarrow P(t) \in OPS^m_{\rho,\delta} \tag{0.9.17}$$

provided

$$\rho > \frac{1}{2}, \quad \delta = 1 - \rho. \tag{0.9.18}$$

One needs $\delta = 1 - \rho$ to insure that $p(\mathcal{C}(t)(x,\xi)) \in S^m_{\rho,\delta}$, and one needs $\rho > \delta$ to insure that the transport equations generate $q_j(t,x,\xi)$ of progressively lower order.

§0.10. Microlocal regularity

We define the notion of wave front set of a distribution $u \in H^{-\infty}(\mathbb{R}^n) = \cup_s H^s(\mathbb{R}^n)$, which refines the notion of singular support. If $p(x,\xi) \in S^m$ has principal symbol $p_m(x,\xi)$, homogeneous in ξ, then the characteristic set of $P = p(x,D)$ is given by

$$\text{Char } P = \{(x,\xi) \in \mathbb{R}^n \times (\mathbb{R}^n \setminus 0) : p_m(x,\xi) = 0\}. \tag{0.10.1}$$

If $p_m(x,\xi)$ is a $K \times K$ matrix, take the determinant. Equivalently, (x_0,ξ_0) is non-characteristic for P, or P is elliptic at (x_0,ξ_0), if $|p(x,\xi)^{-1}| \leq C|\xi|^{-m}$, for (x,ξ) in a small conic neighborhood of (x_0,ξ_0), and $|\xi|$ large. By definition, a conic set is invariant under the dilations $(x,\xi) \mapsto (x,r\xi)$, $r \in (0,\infty)$. The wave front set is defined by

$$WF(u) = \bigcap\{\text{Char } P : P \in OPS^0,\ Pu \in C^\infty\}. \tag{0.10.2}$$

Clearly $WF(u)$ is a closed conic subset of $\mathbb{R}^n \times (\mathbb{R}^n \setminus 0)$. If π is the projection $(x,\xi) \mapsto x$, we have:

PROPOSITION 0.10.A. $\pi(WF(u)) =$ *sing supp* u.

PROOF: If $x_0 \notin$ sing supp u, there is a $\varphi \in C_0^\infty(\mathbb{R}^n)$, $\varphi = 1$ near x_0, such that $\varphi u \in C_0^\infty(\mathbb{R}^n)$. Clearly $(x_0,\xi) \notin$ Char φ for any $\xi \neq 0$, so $\pi(WF(u)) \subset$ sing supp u.

Conversely, if $x_0 \notin \pi(WF(u))$, then for any $\xi \neq 0$ there is a $Q \in OPS^0$ such that $(x_0,\xi) \notin$ Char Q and $Qu \in C^\infty$. Thus we can construct finitely many $Q_j \in OPS^0$ such that $Q_j u \in C^\infty$ and each (x_0,ξ), $|\xi| = 1$ is noncharacteristic for some Q_j. Let $Q = \sum Q_j^* Q_j \in OPS^0$. Then Q is elliptic near x_0 and $Qu \in C^\infty$, so u is C^∞ near x_0.

We define the associated notion of $ES(P)$ for a pseudodifferential operator. Let U be an open conic subset of $\mathbb{R}^n \times (\mathbb{R}^n \setminus 0)$. We say $p(x,\xi) \in S^m_{\rho,\delta}$ has order $-\infty$ on U if for each closed conic set V of U we have estimates, for each N,

$$|D_x^\beta D_\xi^\alpha p(x,\xi)| \leq C_{\alpha\beta NV} \langle\xi\rangle^{-N}, \quad (x,\xi) \in V. \tag{0.10.3}$$

If $P = p(x,D) \in OPS^m_{\rho,\delta}$, we define the essential support of P (and of $p(x,\xi)$) to be the smallest closed conic set on the complement of which $p(x,\xi)$ has order $-\infty$. We denote this set by $ES(P)$.

From the symbol calculus of §0.3 it follows easily that

$$ES(P_1P_2) \subset ES(P_1) \cap ES(P_2) \tag{0.10.4}$$

provided $P_j \in OPS^{m_j}_{\rho_j,\delta_j}$ and $\rho_1 > \delta_2$. To relate $WF(Pu)$ to $WF(u)$ and $ES(P)$, we begin with the following.

LEMMA 0.10.B. *Let $u \in H^{-\infty}(\mathbb{R}^n)$ and suppose U is a conic open set satisfying $WF(u) \cap U = \emptyset$. If $P \in OPS^m_{\rho,\delta}$, $\rho > 0$, and $\delta < 1$, and $ES(P) \subset U$, then $Pu \in C^\infty$.*

PROOF: Taking $P_0 \in OPS^0$ with symbol identically 1 on a conic neighborhood of $ES(P)$, so $P = PP_0 \mod OPS^{-\infty}$, it suffices to conclude that $P_0u \in C^\infty$, so we can specialize the hypothesis to $P \in OPS^0$.

By hypothesis, we can find $Q_j \in OPS^0$ such that $Q_ju \in C^\infty$ and each $(x,\xi) \in ES(P)$ is noncharacteristic for some Q_j, and if $Q = \sum Q_j^*Q_j$, then $Qu \in C^\infty$ and $\text{Char}\, Q \cap ES(P) = \emptyset$. We claim there exists $A \in OPS^0$ such that $AQ = P \mod OPS^{-\infty}$. Indeed, let $\tilde{Q}$ be an elliptic operator whose symbol equals that of Q on a conic neighborhood of $ES(P)$, and let $\tilde{Q}^{-1}$ denote a parametrix for $\tilde{Q}$. Now simply set set $A = P\tilde{Q}^{-1}$. Consequently (mod C^∞) $Pu = AQu \in C^\infty$, so the lemma is proved.

We are ready for the basic result on the preservation of wave front sets by a pseudodifferential operator.

PROPOSITION 0.10.C. *If $u \in H^{-\infty}$ and $P \in OPS^m_{\rho,\delta}$, with $\rho > 0$, $\delta < 1$, then*

$$WF(Pu) \subset WF(u) \cap ES(P). \tag{0.10.5}$$

PROOF: First we show $WF(Pu) \subset ES(P)$. Indeed, if $(x_0,\xi_0) \notin ES(P)$, choose $Q = q(x,D) \in OPS^0$ such that $q(x,\xi) = 1$ on a conic neighborhood of (x_0,ξ_0) and $ES(Q) \cap ES(P) = \emptyset$. Thus $QP \in OPS^{-\infty}$, so $QPu \in C^\infty$. Hence $(x_0,\xi_0) \notin WF(Pu)$.

In order to show that $WF(Pu) \subset WF(u)$, let Γ be any conic neighborhood of $WF(u)$ and write $P = P_1 + P_2$, $P_j \in OPS^m_{\rho,\delta}$, with $ES(P_1) \subset \Gamma$ and $ES(P_2) \cap WF(u) = \emptyset$. By Lemma 0.10.B, $P_2u \in C^\infty$. Thus $WF(u) = WF(P_1u) \subset \Gamma$, which shows $WF(Pu) \subset WF(u)$.

One says that a pseudodifferential operator of type (ρ,δ), with $\rho > 0$ and $\delta < 1$, is microlocal. As a corollary, we have the following sharper form of local regularity for elliptic operators, called microlocal regularity.

COROLLARY 0.10.D. *If $P \in OPS^m_{\rho,\delta}$ is elliptic, $0 \le \delta < \rho \le 1$, then*

$$WF(Pu) = WF(u). \tag{0.10.6}$$

PROOF: We have seen that $WF(Pu) \subset WF(u)$. On the other hand, if $E \in OPS^{-m}_{\rho,\delta}$ is a parametrix for P, we see that $WF(u) = WF(EPu) \subset WF(Pu)$. In fact, by an argument close to the proof of Lemma 0.10.B, we have for general P that

$$WF(u) \subset WF(Pu) \cup \text{Char } P. \tag{0.10.7}$$

We next discuss how the solution operator e^{itA} to a scalar hyperbolic equation $\partial u/\partial t = iA(x,D)u$ propagates the wave front set. We assume $A(x,\xi) \in S^1_{cl}$, with real principal symbol. Suppose $WF(u) = \Sigma$. Then there is a countable family of operators $p_j(x,D) \in OPS^0$, each of whose complete symbols vanish in a neighborhood of Σ, but such that

$$\Sigma = \bigcap_j \{(x,\xi) : p_j(x,\xi) = 0\}. \tag{0.10.8}$$

We know that $p_j(x,D)u \in C^\infty$ for each j. Using Egorov's Theorem, we want to construct a family of pseudodifferential operators $q_j(x,D) \in OPS^0$ such that $q_j(x,D)e^{itA}u \in C^\infty$, this family being rich enough to describe the wave front set of $e^{itA}u$.

Indeed, let $q_j(x,D) = e^{itA}p_j(x,D)e^{-itA}$. Egorov's Theorem implies that $q_j(x,D) \in OPS^0$, (modulo a smoothing operator) and gives the principal symbol of $q_j(x,D)$. Since $p_j(x,D)u \in C^\infty$, we have $e^{itA}p_j(x,D)u \in C^\infty$, which in turn implies $q_j(x,D)e^{itA}u \in C^\infty$. From this it follows that $WF(e^{itA}u)$ is contained in the intersection of the characteristics of the $q_j(x,D)$, which is precisely $C(t)\Sigma$, the image of Σ under the canonical transformation $C(t)$, generated by H_{A_1}. In other words,

$$WF(e^{itA}u) \subset C(t)WF(u).$$

However, our argument is reversible; $u = e^{-itA}(e^{itA}u)$. Consequently, we have:

PROPOSITION 0.10.E. *If $A = A(x,D) \in OPS^1$ is scalar with real principal symbol, then, for $u \in H^{-\infty}$,*

$$WF(e^{itA}u) = C(t)WF(u). \tag{0.10.9}$$

The same argument works for the solution operator $S(t,0)$ to a time-dependent scalar hyperbolic equation.

§0.11. L^p estimates

As shown in §0.2, if $p(x, D) \in OPS^0_{1,\delta}$, $0 \leq \delta \leq 1$, then its Schwartz kernel $K(x, y)$ satisfies estimates

$$|K(x,y)| \leq C|x-y|^{-n}, \tag{0.11.1}$$

and

$$|\nabla_{x,y} K(x,y)| \leq C|x-y|^{-n-1}. \tag{0.11.2}$$

Furthermore, at least when $\delta < 1$, we have an L^2 bound:

$$||Pu||_{L^2} \leq K||u||_{L^2}, \tag{0.11.3}$$

and smoothings of these operators have smooth Schwartz kernels satisfying (0.11.1)-(0.11.3) for fixed C, K. Our main goal here is to sketch a proof of the following fundamental result of Calderon and Zygmund.

THEOREM 0.11.A. *Suppose $P : L^2(\mathbb{R}^n) \to L^2(\mathbb{R}^n)$ is a weak limit of operators with smooth Schwartz kernels satisfying* (0.11.1)-(0.11.3) *uniformly. Then*

$$P : L^p(\mathbb{R}^n) \longrightarrow L^p(\mathbb{R}^n), \quad 1 < p < \infty. \tag{0.11.4}$$

The hypotheses do not imply boundedness on $L^1(\mathbb{R}^n)$ or on $L^\infty(\mathbb{R}^n)$. They will imply that P is of weak type $(1, 1)$. By definition, an operator P is of weak type (q, q) provided that, for any $\lambda > 0$,

$$\text{meas}\,\{x : |Pu(x)| > \lambda\} \leq (C/\lambda^q)||u||^q_{L^q}. \tag{0.11.5}$$

Any bounded operator on L^q is a fortiori of weak type (q, q), in view of the simple inequality

$$\text{meas}\,\{x : |u(x)| > \lambda\} \leq (1/\lambda)||u||_{L^1}. \tag{0.11.6}$$

A key ingredient in proving Theorem 0.11.A is the following result.

PROPOSITION 0.11.B. *Under the hypotheses of Theorem* 0.11.A, *P is of weak type* $(1, 1)$.

Once this is established, Theorem 0.11.A will then follow from the next result, known as the Marcinkiewicz Interpolation Theorem.

PROPOSITION 0.11.C. *If $r < p < q$ and if P is both of weak type (r, r) and of weak type (q, q), then $T : L^p \to L^p$.*

See [S1] for a proof of this result. Also in [S1] is a proof of the following decomposition lemma of Calderon and Zygmund.

LEMMA 0.11.D. *Let $u \in L^1(\mathbb{R}^n)$ and $\lambda > 0$ be given. Then there exist $v, w_k \in L^1(\mathbb{R}^n)$ and disjoint cubes Q_k, $1 \leq k < \infty$, with centers x_k, such that*

$$u = v + \sum_k w_k, \quad ||v||_{L^1} + \sum_k ||w_k||_{L^1} \leq 3||u||_{L^1}, \tag{0.11.7}$$

$$|v(x)| \leq 2^n \lambda, \tag{0.11.8}$$

$$\int_{Q_k} w_k(x)dx = 0 \text{ and supp } w_k \subset Q_k, \tag{0.11.9}$$

$$\sum_k \text{meas}(Q_k) \leq \lambda^{-1}||u||_{L^1}. \tag{0.11.10}$$

One thinks of v as the 'good' piece and $w = \sum w_k$ as the 'bad' piece. What is 'good' about v is that $||v||^2_{L^2} \leq 2^n \lambda ||u||_{L^1}$, so

$$||Pv||^2_{L^2} \leq K^2 ||v||^2_{L^2} \leq 4^n K^2 \lambda ||u||_{L^1}. \tag{0.11.11}$$

Hence

$$(\lambda/2)^2 \text{ meas } \{x : |Pv(x)| > \lambda/2\} \leq C\lambda ||u||_{L^1}. \tag{0.11.12}$$

To treat the action of P on the 'bad' term w, we make use of the following essentially elementary estimate on the Schwartz kernel K.

LEMMA 0.11.E. *There is a $C_0 < \infty$ such that, for any $t > 0$, if $|y| \leq t, x_0 \in \mathbb{R}^n$,*

$$\int_{|x| \geq 2t} \big|K(x, x_0 + y) - K(x, x_0)\big| dx \leq C_0. \tag{0.11.13}$$

To estimate Pw, we have

$$\begin{aligned} Pw_k(x) &= \int K(x,y)w_k(y)dy \\ &= \int_{Q_k} \left[K(x,y) - K(x,x_k)\right] w_k(y)dy. \end{aligned} \tag{0.11.14}$$

Before we make further use of this, a little notation: Let Q_k^* be the cube concentric with Q_k, enlarged by a linear factor of $2n^{\frac{1}{2}}$, so meas $Q_k^* = (4n)^{n/2}$ meas Q_k. For some $t_k > 0$, we can arrange that

$$\begin{aligned} &Q_k \subset \{x : |x - x_k| \le t_k\} \\ &Y_k = \mathbb{R}^n \setminus Q_k^* \subset \{x : |x - x_k| > 2t_k\}. \end{aligned} \tag{0.11.15}$$

Furthermore, set $\mathcal{O} = \cup Q_k^*$, and note that

$$\text{meas } \mathcal{O} \le (L/\lambda)\|u\|_{L^1} \tag{0.11.16}$$

with $L = (4n)^{n/2}$. Now, from (0.11.14), we have

(0.11.17)

$$\begin{aligned} &\int_{Y_k} |Pw_k(x)|dx \\ &\quad\le \int_{|y|\le t_k,} \int_{|x|\ge 2t_k} \left|K(x+x_k,y) - K(x+x_k,x_k)\right| \cdot |w_k(y+x_k)|dx\ dy \\ &\quad\le C_0\|w_k\|_{L^1}, \end{aligned}$$

the last estimate using Lemma 0.11.E. Thus

$$\int_{\mathbb{R}^n\setminus\mathcal{O}} |Pw(x)|dx \le 3C_0\|u\|_{L^1}. \tag{0.11.18}$$

Together with (0.11.16), this gives

$$(\lambda/2) \text{ meas } \{x : |Pw(x)| > \lambda/2\} \le C_1\|u\|_{L^1}, \tag{0.11.19}$$

and this estimate together with (0.11.12) yields the desired weak (1,1) estimate:

$$\text{meas } \{x : |Pu(x)| > \lambda\} \le (C_2/\lambda)\|u\|_{L^1}. \tag{0.11.20}$$

This completes the proof.

We next describe an important generalization to operators acting on Hilbert space valued functions. Let $\mathcal{H}_1$ and $\mathcal{H}_2$ be Hilbert spaces and suppose

$$P : L^2(\mathbb{R}^n, \mathcal{H}_1) \longrightarrow L^2(\mathbb{R}^n, \mathcal{H}_2). \tag{0.11.21}$$

The P has an $\mathcal{L}(\mathcal{H}_1, \mathcal{H}_2)$-operator valued Schwartz kernel K. Let us impose on K the hypotheses of Theorem 0.11.A, where now $|K(x, y)|$ stands for the $\mathcal{L}(\mathcal{H}_1, \mathcal{H}_2)$-norm of $K(x, y)$. Then all the steps in the proof of Theorem 0.11.A extend to this case. Rather than formally state this general result, we will concentrate on an important special case.

PROPOSITION 0.11.F. *Let $P(\xi) \in C^\infty(\mathbb{R}^n, \mathcal{L}(\mathcal{H}_1, \mathcal{H}_2))$ satisfy*

$$\|D_\xi^\alpha P(\xi)\|_{\mathcal{L}(\mathcal{H}_1,\mathcal{H}_2)} \le C_\alpha \langle \xi \rangle^{-|\alpha|} \tag{0.11.22}$$

for all $\alpha \ge 0$. Then

$$P(D) : L^p(\mathbb{R}^n, \mathcal{H}_1) \longrightarrow L^p(\mathbb{R}^n, \mathcal{H}_2) \text{ for } 1 < p < \infty. \tag{0.11.23}$$

This leads to an important circle of results known as *Littlewood-Paley Theory*. To obtain this, start with a partition of unity

$$1 = \sum_{j=0}^{\infty} \varphi_j(\xi)^2 \tag{0.11.24}$$

where $\varphi_j \in C^\infty$, $\varphi_0(\xi)$ is supported on $|\xi| \le 1$, $\varphi_1(\xi)$ is supported on $\frac{1}{2} \le |\xi| \le 2$, and $\varphi_j(\xi) = \varphi_1(2^{1-j}\xi)$ for $j \ge 2$. We take $\mathcal{H}_1 = \mathbb{C}$, $\mathcal{H}_2 = \ell^2$, and look at

$$\Phi : L^2(\mathbb{R}^n) \longrightarrow L^2(\mathbb{R}^n, \ell^2) \tag{0.11.25}$$

given by

$$\Phi(f) = \big(\varphi_0(D)f, \varphi_1(D)f, \varphi_2(D)f, \dots\big). \tag{0.11.26}$$

This is clearly an isometry, though of course it is not surjective. The adjoint

$$\Phi^* : L^2(\mathbb{R}^n, \ell^2) \longrightarrow L^2(\mathbb{R}^n), \tag{0.11.27}$$

given by

$$\Phi^*(g_0, g_1, g_2, \dots) = \sum \varphi_j(D) g_j \tag{0.11.28}$$

satisfies

$$\Phi^*\Phi = I \tag{0.11.29}$$

on $L^2(\mathbb{R}^n)$. Note that $\Phi = \Phi(D)$, where

$$\Phi(\xi) = (\varphi_0(\xi), \varphi_1(\xi), \varphi_2(\xi), \dots). \tag{0.11.29}$$

It is easy to see that the hypthesis (0.11.22) is satisfied by both $\Phi(\xi)$ and $\Phi^*(\xi)$. Hence, for $1 < p < \infty$,

$$\begin{aligned} \Phi &: L^p(\mathbb{R}^n) \longrightarrow L^p(\mathbb{R}^n, \ell^2) \\ \Phi^* &: L^p(\mathbb{R}^n, \ell^2) \longrightarrow L^p(\mathbb{R}^n). \end{aligned} \tag{0.11.30}$$

In particular, Φ maps $L^p(\mathbb{R}^n)$ isomorphically onto a closed subspace of $L^p(\mathbb{R}^n, \ell^2)$, and we have compatibility of norms:

$$\|u\|_{L^p} \approx \|\Phi u\|_{L^p(\mathbb{R}^n, \ell^2)}. \tag{0.11.31}$$

In other words,

$$C_p'\|u\|_{L^p} \le \Big\| \sum_{j=0}^{\infty} |\varphi_j(D)u|^2 \Big\|_{L^p}^{\frac{1}{2}} \le C_p\|u\|_{L^p}, \tag{0.11.32}$$

for $1 < p < \infty$. This Littlewood-Paley estimate will be used in Chapter 2.

§0.12. Operators on manifolds

If M is a smooth manifold, a continuous linear operator $P : C_0^\infty(M) \to \mathcal{D}'(M)$ is said to be a pseudodifferential operator in $OPS^m_{\rho,\delta}(M)$ provided its Schwartz kernel is C^∞ off the diagonal in $M \times M$, and there exists an open cover Ω_j of M, a subordinate partition of unity φ_j, and diffeomorphisms $F_j : \Omega_j \to \mathcal{O}_j \subset \mathbb{R}^n$ which transform the operators $\varphi_k P \varphi_j : C^\infty(\Omega_j) \to \mathcal{E}'(\Omega_k)$ into pseudodifferential operators in $OPS^m_{\rho,\delta}$, as defined in §0.1.

This is a rather 'liberal' definition of $OPS^m_{\rho,\delta}(M)$. For example, it poses no growth restrictions on the Schwartz kernel $K \in \mathcal{D}'(M \times M)$ at infinity. Consequently, if M happens to be $\mathbb{R}^n$, the class of operators in $OPS^m_{\rho,\delta}(M)$ as defined above is a bit larger than the class $OPS^m_{\rho,\delta}$ defined in §0.1. One negative consequence of this definition is that pseudodifferential operators cannot always be composed. One drastic step to fix this would be to insist that the kernel be properly supported, so $P : C_0^\infty(M) \to C_0^\infty(M)$. If M is compact, these problems do not arise. If M is noncompact, it is often of interest to place specific restrictions on K near infinity, but analytical problems inducing one to do so will not be studied in this paper.

Another way in which the definition of $OPS^m_{\rho,\delta}(M)$ given above is liberal is that it requires P to be locally transformed to pseudodifferential operators on $\mathbb{R}^n$ by *some* coordinate cover. One might ask if then P is necessarily so transformed by *every* coordinate cover. This comes down to asking if the class $OPS^m_{\rho,\delta}$ defined in §0.1 is invariant under a coordinate transformation, i.e., a diffeomorphism $F : \mathbb{R}^n \to \mathbb{R}^n$. It would suffice to establish this for the case where F is the identity outside a compact set.

In case $\rho \in (\frac{1}{2}, 1]$ and $\delta = 1 - \rho$, this invariance is a special case of the Egorov Theorem established in §0.9. Indeed, one can find a time-dependent vector field $X(t)$ whose flow at $t = 1$ coincides with F and apply Theorem 0.9.A to $iA(t, x, D) = X(t)$. Note that the formula for the principal symbol of the conjugated operator given there implies

$$p(1, F(x), \xi) = p_0(x, F'(x)^t \xi), \tag{0.12.1}$$

so that the principal symbol is well defined on the *cotangent bundle* of M.

Alternatively, one can insert the coordinate changes into the Fourier integral representation of P and work on that. This latter approach has the advantage of working for a larger set of symbol classes $S^m_{\rho,\delta}$ than the more general conjugation invariance applies to. In fact, one needs only

$$\rho > \frac{1}{2}, \quad \rho + \delta \geq 1.$$

A proof of this can be found in [H1], [T2]. While this coordinate invariance is good to know, it will not play a crucial role in the analysis done in this paper.

Chapter 1: Symbols with limited smoothness

Here we establish some very general facts about symbols $p(x,\xi)$ with limited smoothness in x. We prove some operator bounds on $p(x,D)$ when $p(x,\xi)$ is homogeneous in ξ.

In §1.2 we show that these simple results lead to some easy regularity theorems, for solutions to an elliptic PDE, $F(x, D^m u) = f$, under the assumption that u already possesses considerable smoothness, e.g., roughly $2m$ derivatives. Though this is a rather weak result, which will be vastly improved in Chapters 2 and 3, nevertheless it has some uses, beyond providing a preliminary example of techniques to be developed here. For example, when examining local solvability of $F(x, D^m u) = f$, one can use a Banach space implicit function theorem to find $u \in H^s$ with s large, and then apply such a regularity result as Theorem 1.2.D to obtain local C^∞ solutions.

One key tool for further use of symbols introduced here is to write $p(x,\xi) = p^\#(x,\xi) + p^b(x,\xi)$, with $p^\#(x,\xi)$ a C^∞ symbol, in $S^m_{1,\delta}$, and $p^b(x,\xi)$ having lower order. This symbol decomposition is studied in §1.3.

§1.1. Symbol classes

We introduce here some general classes of symbols $p(x,\xi)$ which have limited regularity in x. To start with, let X be any Banach space of functions, such that

$$C_0^\infty \subset X \subset C^0. \tag{1.1.1}$$

We say

$$p(x,\xi) \in XS^m_{1,0} \Longleftrightarrow \|D_\xi^\alpha p(\cdot,\xi)\|_X \le C_\alpha \langle\xi\rangle^{m-|\alpha|}, \ \alpha \ge 0. \tag{1.1.2}$$

For applications, we will generally want X to be a Banach algebra under pointwise multiplication, and more specifically

$$\begin{gathered} f \in C^\infty(\mathbb{R}), \ u \in X \Longrightarrow f(u) \in X; \\ f \text{ maps bounded sets to bounded sets in } X. \end{gathered} \tag{1.1.3}$$

Such an X as we consider will usually be one of a family $\{X^s : s \in \Sigma\}$ of spaces, known as a scale. The set Σ will be of the form $[\sigma_0, \infty)$ or (σ_0, ∞), and we assume

$$X^s \subset X^t \text{ if } t < s, \tag{1.1.4}$$

provided $t, s \in \Sigma$, and, if $s \in \Sigma, m \in \mathbb{Z}^+$, then $s + m \in \Sigma$ and

$$OPD^m : X^{s+m} \longrightarrow X^s, \tag{1.1.5}$$

where OPD^m denotes the space of differential operators of order m (with smooth coefficients).

Examples of scales satisfying the conditions above are

$$X^s = C^s(\mathbb{R}^n), \quad \Sigma = [0, \infty) \tag{1.1.6}$$

and

$$X^s = H^{s,p}(\mathbb{R}^n), \quad \Sigma = (n/p, \infty), \tag{1.1.7}$$

for any given $p \in [1, \infty)$. In (1.1.6), C^s is the space of C^k functions whose k^{th} derivatives satisfy the Hölder condition

$$|u(x+y) - u(x)| \leq C|y|^\sigma, \quad |y| \leq 1, \tag{1.1.8}$$

where $s = k + \sigma$, $0 \leq \sigma < 1$. The spaces (1.1.7) are Sobolev spaces.

We say $\{X^s\}$ is microlocalizable if, for $m \in \mathbb{R}$, $s, s+m \in \Sigma$,

$$OPS^m_{1,0} : X^{s+m} \longrightarrow X^s. \tag{1.1.9}$$

The Sobolev spaces (1.1.7) have this property provided $p \in (1, \infty)$. The property (1.1.9) fails for the spaces C^s if s is an integer. In such a case, one needs to use the Zygmund spaces

$$X^s = C^s_*(\mathbb{R}^n), \quad \Sigma = (0, \infty), \tag{1.1.10}$$

which coincide with C^s is s is not an integer, but differ from C^s if s is an integer. Some important properties of Sobolev spaces and Zygmund spaces are discussed in Appendix A.

We will say $p(x, \xi) \in XS^m_{cl}$, or merely XS^m, provided $p(x, \xi) \in XS^m_{1,0}$ and $p(x, \xi)$ has a classical expansion

$$p(x, \xi) \sim \sum_{j \geq 0} p_j(x, \xi) \tag{1.1.11}$$

in terms homogeneous of degree $m - j$ in ξ (for $|\xi| \geq 1$), in the sense that the difference between $p(x, \xi)$ and the sum over $j < N$ belongs to $XS^{m-N}_{1,0}$.

As usual, we define the operator associated to $p(x, \xi)$

$$p(x, D)u = \int p(x, \xi)\, \hat{u}(\xi)\, e^{ix \cdot \xi}\, d\xi. \tag{1.1.12}$$

We also consider $p(D, x)$, defined by

$$p(D, x)u = (2\pi)^{-n} \iint p(y, \xi)\, e^{i(x-y)\cdot \xi}\, u(y)\, dy\, d\xi. \tag{1.1.13}$$

We now derive some mapping properties for the operators (1.1.12) - (1.1.13), in the case $p(x,\xi) \in X^s S^m_{cl}$. The analogues for $p(x,\xi) \in X^s S^m_{1,0}$ are somewhat harder to establish. These will be discussed in §2.1.

PROPOSITION 1.1.A. *Assume $\{X^s\}$ is a microlocalizable scale. If $m \in \mathbb{R}$, $s \in \Sigma$, $p(x,\xi) \in X^s S^m_{cl}$, then*

$$(1.1.14) \qquad p(x,D) : X^{s+m} \longrightarrow X^s \text{ if } s+m \in \Sigma,$$

and

$$(1.1.15) \qquad p(D,x) : X^s \longrightarrow X^{s-m} \text{ if } s-m \in \Sigma.$$

PROOF: Considering $p(x,D)(1-\Delta)^{-\frac{1}{2}m}$ and $(1-\Delta)^{-\frac{1}{2}m}p(D,x)$, respectively, it suffices to treat the case $m = 0$. Via such decompositions as

$$(1.1.16) \qquad p_0(x,\xi) = \sum_{l=0}^{\infty} p_{0l}(x)\omega_l(\xi/|\xi|),$$

where $\omega_l \in C^\infty(S^{n-1})$ are the spherical harmonics, such mapping properties are easily established. The operators $\omega_l(D)$ have norms bounded by $C\langle l\rangle^K$, if hypothesis (1.1.9) holds, while the factors p_{0l} have rapidly decreasing norms as $l \longrightarrow \infty$. Note that $p_0(D,x)u = \sum_l \omega_l(D)(p_{0l}u)$.

It is worthwhile to record the following generalization. If X and Y are two Banach spaces satisfying (1.1.1), we say X is a Y-module if pointwise multiplication gives a continuous bilinear map $Y \times X \longrightarrow X$.

PROPOSITION 1.1.B. *If $\{X^s\}$ is microlocalizable and $p(x,\xi) \in YS^m_{cl}$, then* (1.1.14) - (1.1.15) *hold, provided X^s is a Y-module.*

In fact, it is sometimes useful to consider a scale $\{X^s\}$ for s in a larger interval $\tilde{\Sigma}$ (containing Σ), on which (1.1.1) might not hold. For example we can consider $X^s = H^{s,p}(\mathbb{R}^n)$, for $s \in \mathbb{R}$. In this more general situation, if Y is a Banach space of functions satisfying (1.1.1) and X^s is a Banach space of distributions, we say X^s is a Y-module provided there is a natural continuous product $Y \times X^s \longrightarrow X^s$. It is clear that Proposition 1.1.B generalizes to this case.

It is occasionally even useful to consider $p(x,\xi) \in YS^m_{1,0}$ for a Banach space Y of distributions not satisfying (1.1.1). For such a case, one would tend to have $m < 0$ and consider only $p(D,x)$. The following result follows in the same way as Proposition 1.1.A and Proposition 1.1.B.

PROPOSITION 1.1.C. *Let X^s, Y, Z and W be Banach spaces of distributions, and assume $p(x,\xi) \in YS^m_{cl}$. Then*

$$p(D,x) : Z \longrightarrow X^s \tag{1.1.17}$$

provided pointwise multiplication yields a continuous bilinear map

$$Y \times Z \longrightarrow W \tag{1.1.18}$$

and

$$OPS^m_{cl} : W \longrightarrow X^s. \tag{1.1.19}$$

A basic family of special cases of (1.1.18) is

$$H^{-s,p}(\mathbb{R}^n) \times H^{r,q}(\mathbb{R}^n) \longrightarrow H^{-s,p}(\mathbb{R}^n), \tag{1.1.20}$$

with

$$\frac{1}{p} + \frac{1}{q} = 1, \quad r > \frac{n}{q}, \quad 0 \le s \le r, \tag{1.1.21}$$

which follows by duality from $H^{s,q} \times H^{r,q} \longrightarrow H^{s,q}$.

§1.2. Some simple elliptic regularity theorems

Throughout this section we suppose $\{X^s : s \in \Sigma\}$ is a microlocalizable scale.

Suppose we have an elliptic differential operator of order m,

$$A(x,D) = \sum_{|\alpha| \le m} a_\alpha(x)\ D^\alpha,$$

with coefficients in X^s, where s is an element of Σ. Then we can take $p(x,\xi) \in X^s S^{-m}_{cl}$, equal to $A(x,\xi)^{-1}$ for $|\xi|$ large. The formal transpose of $A(x,D)$ is given by

$$A(x,D)^t u = \sum_{|\alpha| \le m} (-D)^\alpha a_\alpha(x) u, \tag{1.2.1}$$

and we have

$$\begin{aligned} p(D,x)A(x,D)u &= \iint p(y,\xi)[A(y,D)u(y)]e^{i(x-y)\cdot\xi}\,dy\,d\xi \\ &= \iint u(y)\ A(y,D)^t\left[p(y,\xi)e^{i(x-y)\cdot\xi}\right]dy\,d\xi \\ &= u + Ru \end{aligned} \tag{1.2.2}$$

where

$$Ru(x) = \iint u(y)R(y,\xi)e^{i(x-y)\cdot\xi}dy\, d\xi. \tag{1.2.3}$$

Here

$$\begin{aligned} R(y,\xi) &= e^{iy\cdot\xi}A(y,D)^t\left[p(y,\xi)e^{-iy\cdot\xi}\right] - 1 \\ &= \sum_{j=1}^{m} r_j(y,\xi), \end{aligned} \tag{1.2.4}$$

with

$$r_j(y,\xi) \in X^{s-j}S_{cl}^{-j}. \tag{1.2.5}$$

From (1.1.15) of Proposition 1.1.A, it follows that

$$r_j(D,x) : X^{s-j} \longrightarrow X^s, \text{ if } s-j \in \Sigma, \tag{1.2.6}$$

so

$$R : X^{s-1} \longrightarrow X^s, \text{ if } s-m \in \Sigma. \tag{1.2.7}$$

Now if

$$A(x,D)u = f \tag{1.2.8}$$

then

$$u = p(D,x)f - Ru. \tag{1.2.9}$$

Since $p(D,x) : X^r \longrightarrow X^{r+m}$ for $r \leq s$, we have most of the proof of the following regularity result.

THEOREM 1.2.A. *Let A(x,D) be an elliptic differential operator with coefficients in X^s; assume $s-m \in \Sigma$. Assume $A(x,D)u = f \in X^r$, and assume a priori that $u \in X^{s-1}$. Then $u \in X^{r+m}$, provided $r = s-m+k$, $k = 0,1,\dots,m$*

PROOF: A look at (1.2.9) shows that $u \in X^s$ under these hypotheses, which covers the case $r = s-m$ of the theorem. We now treat the cases $r = s-m+k$, $k = 1,\dots,m$. Thus we suppose $f \in X^{s-m+k}$. By induction, we can assume $u \in X^{s+k-1}$.

Let $u_\beta = D^\beta u$. If $|\beta| \le k$, applying D^β to (1.2.8) gives

$$\sum_{|\alpha|\le m} D^\alpha u_\beta = D^\beta f - \sum_{|\alpha|\le m} \binom{\beta}{\sigma}[D^\sigma a_\alpha(x)]D^{\alpha+\gamma}u \tag{1.2.10}$$
$$= f_\beta \in X^{s-m},$$

where in the second sum, $\sigma+\gamma=\beta$, $|\gamma|\le k-1$. Now we can regard

$$A(x,D)u_\beta = F_\beta \in X^{s-m},\ |\beta|\le k, \tag{1.2.11}$$

as an elliptic system, to which the $k=0$ case of our theorem applies, since $u\in X^{s+k-1} \Longrightarrow u_\beta \in X^{s-1}$. This completes our inductive step and proves the rest of the theorem.

Though it is desirable to have results for less regular coefficients, nevertheless Theorem 1.2.A leads to some useful regularity results for solutions to nonlinear elliptic PDE, as we now show. Consider the equation

$$\sum_{|\alpha|\le m} a_\alpha(x, D^{m-1}u)D^\alpha u = f. \tag{1.2.12}$$

Assume it is elliptic of order m, the coefficients $a_\alpha(x,\zeta)$ being smooth in their arguments.

Theorem 1.2.B. *Assume (1.2.12) holds with $f\in X^s$; assume $u\in X^{s+m-1}$, so $a_\alpha(x,D^{m-1}u)\in X^s$, and assume $s-m\in\Sigma$. Then $u\in X^{s+m}$.*

Proof: With $a_\alpha(x) = a_\alpha(x,D^{m-1}u)$, the case $r=s$ of Theorem 1.2.A applies.

Corollary 1.2.C. *If $u\in X^{s+m-1}$ solves (1.2.12) with $f\in C^\infty$, and $s-m\in\Sigma$, then $u\in C^\infty$.*

We also note the following regularity result for solutions to a completely nonlinear PDE

$$F(x, D^m u) = f, \tag{1.2.13}$$

assumed to be elliptic, $F = F(x,\zeta)$ being smooth in its arguments, $\zeta = (\zeta_\alpha : |\alpha|\le m)$.

Theorem 1.2.D. *Assume $u\in X^{s+m}$ solves (1.2.13), with $f\in X^{s+1}$, and that (1.2.13) is elliptic. Then $u\in X^{s+m+1}$, provided $s-m\in\Sigma$.*

PROOF: Set $u_j = \partial u/\partial x_j$. Then differentiating (1.2.13) with respect to x_j gives

$$(1.2.14)\quad \sum_{|\alpha|\le m} (\partial F/\partial \zeta_\alpha)(x, D^m u) D^\alpha u_j = -F_{x_j}(x, D^m u) + \partial f/\partial x_j$$
$$= f_j \in X^s.$$

As a PDE for u_j, this has similar structure to the quasi-linear PDE (1.2.12), and the analysis proving Theorem 1.2.B applies.

Clearly Corollary 1.2.C has an analogue in this case, when one assumes a priori that $u \in X^{s+m}$.

The main drawback of these results is the hypothesis that $s - m \in \Sigma$, which requires that $s > m+\frac{n}{p}$ in case (1.1.7) and $s > m$ in case (1.1.10). We will obtain sharper results which include the classical results of Schauder in §2.2, after developing stronger tools on §1.3 - §2.1. Here we note a simple improvement that can be made if we use Proposition 1.1.C to tighten up the reasoning leading to (1.2.7).

Namely, suppose the scale $\{X^s : s \in \Sigma\}$, satisfying (1.1.1), (1.1.3), (1.1.9), is enlarged to $\{X^s : s \in \tilde{\Sigma}\}$, satisfying (1.1.9). To get (1.2.7), one really needs in (1.2.6) only that

$$r_j(D, X) : X^{s-1} \longrightarrow X^s, \text{ for } r_j(x, \xi) \in X^{s-j} S^{-j}_{cl}.$$

In some cases one can get this with $s - j \in \tilde{\Sigma} \setminus \Sigma$. As noted in Proposition 1.1.C, this happens if multiplication gives a continuous bilinear map

$$(1.2.15)\qquad X^{s-j} \times X^{s-1} \longrightarrow X^{s-j}, \quad j = 1, \dots, m.$$

For example, if $X^s = H^s(\mathbb{R}^n)$, this holds as long as $s > \frac{1}{2}n + 1$ and $s \ge \frac{1}{2}(m+1)$. This is an improvement over Theorem 1.2.A, which would require $s > \frac{1}{2}n + m$. Nevertheless, it is not nearly as good an improvement as will be obtained in §2.2.

§1.3. Symbol smoothing

Our goal in this section is to write a symbol $p(x,\xi) \in X^s S^m_{1,0}$ as a sum of a smooth symbol and a remainder of lower order. The smooth part will not belong to $S^m_{1,0}$, but rather to one of Hörmander's classes $S^m_{1,\delta}$.

We will use a partition of unity

$$(1.3.1)\qquad 1 = \sum_{j=0}^{\infty} \psi_j(\xi), \quad \psi_j \text{ supported on } \langle \xi \rangle \sim 2^j,$$

such that $\psi_j(\xi) = \psi_1(2^{1-j}\xi)$ for $j \geq 2$. To get this, you can start with positive $\psi_0(\xi)$, equal to 1 for $|\xi| \leq 1, 0$ for $|\xi| \geq 2$, set $\Psi_j(\xi) = \psi_0(2^{-j}\xi)$, and set $\psi_j(\xi) = \Psi_j(\xi) - \Psi_{j-1}(\xi)$ for $j \geq 1$. We will call this an S_1^0 partition of unity. It is also sometimes called a Littlewood-Paley partition of unity.

Given $p(x,\xi) \in X^s S_{1,0}^m$, choose $\delta \in (0,1]$ and set

$$p^{\#}(x,\xi) = \sum_{j=0}^{\infty} J_{\epsilon_j}\, p(x,\xi)\psi_j(\xi) \tag{1.3.2}$$

where J_ϵ is a smoothing operator on functions of x, namely

$$J_\epsilon f(x) = \phi(\epsilon D) f(x) \tag{1.3.3}$$

with $\phi \in C_0^\infty(\mathbb{R}^n)$, $\phi(\xi) = 1$ for $|\xi| \leq 1$ (e.g., $\phi = \psi_0$), and we take

$$\epsilon_j = 2^{-j\delta}. \tag{1.3.4}$$

We then define $p^b(x,\xi)$ to be $p(x,\xi) - p^{\#}(x,\xi)$, so our decomposition is

$$p(x,\xi) = p^{\#}(x,\xi) + p^b(x,\xi). \tag{1.3.5}$$

To analyze these terms, we use the folowing simple result.

LEMMA 1.3.A. *If $\{X^s : s \in \Sigma\}$ is a microlocalizable scale, then, for $\epsilon \in (0,1]$,*

$$||D_x^\beta J_\epsilon f||_{X^s} \leq C_\beta\, \epsilon^{-|\beta|} ||f||_{X^s} \tag{1.3.6}$$

and

$$||f - J_\epsilon f||_{X^{s-t}} \leq C\epsilon^t ||f||_{X^s} \text{ for } s, s-t \in \Sigma,\ t \geq 0. \tag{1.3.7}$$

PROOF: The estimate (1.3.6) follows from the fact that, for each $\beta \geq 0$,

$$\epsilon^{|\beta|} D_x^\beta \phi(\epsilon D) \text{ is bounded in } OPS_{1,0}^0$$

and the estimate (1.3.7) follows from the fact that, with $\Lambda = (1-\Delta)^{\frac{1}{2}}$,

$$\Lambda^t : X^s \longrightarrow X^{s-t} \text{ isomorphically, if } s, s-t \in \Sigma,$$

plus the fact that

$$\epsilon^{-t}\Lambda^{-t}(1 - \phi(\epsilon D)) \text{ is bounded in } OPS_{1,0}^0,$$

for $0 < \epsilon \le 1$.

Using this, we easily derive the following conclusion.

PROPOSITION 1.3.B. *If $\{X^s\}$ is a microlocalizable scale and $p(x,\xi) \in X^s S^m_{1,0}$, then, with the decomposition (1.3.5) defined by (1.3.2) - (1.3.4), we have*

$$p^{\#}(x,\xi) \in S^m_{1,\delta} \tag{1.3.8}$$

and

$$p^b(x,\xi) \in X^{s-t} S^{m-t\delta}_{1,0} \text{ if } s, s-t \in \Sigma. \tag{1.3.9}$$

PROOF: The estimate (1.3.6) yields

$$\|D_x^\beta D_\xi^\alpha p^{\#}(\cdot,\xi)\|_{X^s} \le C_{\alpha\beta}\langle\xi\rangle^{m-|\alpha|+\delta|\beta|}, \tag{1.3.10}$$

which implies (1.3.8) since $X^s \in C^0$. Similarly (1.3.9) follows from (1.3.7).

A primary class of symbols we will deal with is $C^s S^m_{1,0}$, equal to $C^s_* S^m_{1,0}$ for $s \in \mathbb{R}^+ \setminus \mathbb{Z}^+$, and contained in the latter for $s \in \mathbb{Z}^+$. We can obtain more precise results on the decomposition of $p(x,\xi) \in C^s S^m_{1,0}$ by the following supplement to Lemma 1.3.A.

LEMMA 1.3.C. *Given $f \in C^s$, $s > 0$, we have*

$$\|D_x^\beta J_\epsilon f\|_{L^\infty} \le \begin{cases} C\|f\|_{C^s}, & |\beta| \le s, \\ C\epsilon^{-(|\beta|-s)}\|f\|_{C^s_*}, & |\beta| > s, \end{cases} \tag{1.3.11}$$

and

$$\|f - J_\epsilon f\|_{L^\infty} \le C_s \epsilon^s \|f\|_{C^s_*}. \tag{1.3.12}$$

PROOF: The first estimate in (1.3.11) (for $|\beta| \le s$) is trivial. If $|\beta| > s$, we have, for $\epsilon \sim 2^{-j}$,

$$\begin{aligned} \|D^\beta \phi(\epsilon D) f\|_{L^\infty} &\le \sum_{l \le j} \|D^\beta \psi_l(D) f\|_{L^\infty} \\ &\le C \sum_{l \le j} 2^{l|\beta|} \|\psi_l(D) f\|_{L^\infty} \\ &\le C \sum_{l \le j} 2^{l|\beta|} \cdot 2^{-ls} \|f\|_{C^s_*}. \end{aligned} \tag{1.3.13}$$

Since $\sum_{l\leq j} 2^{l(|\beta|-s)} \leq C_s 2^{j(|\beta|-s)}$ for $s < |\beta|$, we have the rest of (1.3.11). To obtain (1.3.12), if $\epsilon \sim 2^{-j}$, we have

$$\|(1-\phi(\epsilon D))f\|_{L^\infty} \leq \sum_{l\geq j} \|\psi_l(D)f\|_{L^\infty} \leq C \sum_{l\geq j} 2^{-ls}\|f\|_{C^s_*}, \tag{1.3.14}$$

and since $\sum_{l\geq j} 2^{-ls} \leq C_s 2^{-js}$ for $s > 0$, we have (1.3.12).

Exploiting (1.3.11) gives the following improvement of (1.3.8).

PROPOSITION 1.3.D. *If $p(x,\xi) \in C^s S^m_{1,0}$ has decomposition (1.3.5), then*

$$D^\beta_x p^\#(x,\xi) \in \begin{cases} S^m_{1,\delta} & \text{for } |\beta| \leq s, \\ S^{m+\delta(|\beta|-s)}_{1,\delta} & \text{for } |\beta| > s. \end{cases} \tag{1.3.15}$$

In the course of using the decomposition(1.3.5), we will find it helpful to have the following generalization of $X^s S^m_{1,0}$, at least for $X^s = C^s$ or C^s_*. For $\delta \in [0,1]$, we say $p(x,\xi)$ belongs to $C^s_* S^m_{1,\delta}$ provided

$$|D^\alpha_\xi p(x,\xi)| \leq C_\alpha \langle\xi\rangle^{m-|\alpha|} \tag{1.3.16}$$

and

$$\|D^\alpha_\xi p(\cdot,\xi)\|_{C^s_*} \leq C_\alpha \langle\xi\rangle^{m-|\alpha|+s\delta}. \tag{1.3.17}$$

We will say $p(x,\xi) \in C^s S^m_{1,\delta}$ if, in addition

$$\|D^\alpha_\xi p(\cdot,\xi)\|_{C^j} \leq C_\alpha \langle\xi\rangle^{m-|\alpha|+j\delta} \text{ for } 0 \leq j \leq s. \tag{1.3.18}$$

Thus we make a semantic distinction between $C^s_* S^m_{1,\delta}$ and $C^s S^m_{1,\delta}$, even when $s \notin \mathbb{Z}^+$, in which case C^s_* and C^s coincide.

PROPOSITION 1.3.E. *If $p(x,\xi) \in C^s S^m_{1,0}$, then, in the decomposition (1.3.5),*

$$p^b(x,\xi) \in C^s S^{m-s\delta}_{1,\delta}. \tag{1.3.19}$$

PROOF: That $p^b(x,\xi)$ satisfies an estimate of the form (1.3.17), with m replaced by $m - s\delta$, follows from (1.3.7), with $t = 0$. That, for integer j, $0 \leq j \leq s$, we have an estimate of the form (1.3.18), with m replaced by $m - s\delta$, follows from (1.3.12) and its easy generalization

$$\|f - J_\epsilon f\|_{C^j} \leq C\, \epsilon^{s-j}\|f\|_{C^s_*}, \quad 0 \leq j < s, \tag{1.3.20}$$

and from the simple estimate $\|f - J_\epsilon f\|_{C^j} \le C\|f\|_{C^j}$, in case $s = j$ is an integer (which in turn follows from the first part of (1.3.11)).

It will also occasionally be useful to smooth out a symbol $p(x,\xi) \in C^s S^m_{1,\delta}$, for $\delta \in (0,1)$. Pick $\gamma \in (\delta, 1)$ and apply (1.3.2), with $\epsilon_j = 2^{-j(\gamma-\delta)}$, obtaining $p^\#(x,\xi)$ and hence a decomposition of the form (1.3.5). In this case, we obtain

$$p(x,\xi) \in C^s S^m_{1,\delta} \Longrightarrow p^\#(x,\xi) \in S^m_{1,\gamma},\ p^b(x,\xi) \in C^s S^{m-(\gamma-\delta)s}_{1,\gamma}. \tag{1.3.21}$$

Chapter 2: Operator estimates and elliptic regularity

In order to make use of the symbol smoothing of §1.3, we need operator estimates on $p(x,D)$ when $p(x,\xi) \in C^s S^m_{1,\delta}$. We give a number of such results in §2.1, most of them following from work of Bourdaud [BG]. The main result, Theorem 2.1.A, treats the case $\delta = 1$. This will be very useful for the treatment of paradifferential operators in Chapter 3.

In §2.2 we derive a number of regularity results for nonlinear elliptic PDE, including the classical Schauder estimates, and some variants. Section 2.3 gives a brief treatment of adjoints of operators with symbols in $C^r S^1_{1,0}$. In §2.4 we prove a sharp Gårding inequality for $p(x,\xi) \in C^r S^m_{1,0}$. We use the symbol decomposition $p(x,\xi) = p^\#(x,\xi) + p^b(x,\xi)$ in such a manner that we can apply the Fefferman-Phong inequality to $p^\#(x,D)$ and crudely bound $p^b(x,D)$. We also establish an 'ordinary' Gårding inequality which will be useful.

§2.1. Bounds for operators with nonregular symbols

It is more difficult to establish continuity of operators with symbols in $X^s S^m_{1,0}$, and other classes which arose in Chapter 1, than those with symbols in $X^s S^m_{cl}$, but very important to do so, in order to exploit the symbol smoothing of §1.3. Most of our continuity results will be consequences of the following result of G. Bourdaud [BG], itself following pioneering work of Stein [S2].

THEOREM 2.1.A. *If $r > 0$ and $p \in (1,\infty)$, then, for $p(x,\xi) \in C^r_* S^m_{1,1}$,*

$$p(x,D) : H^{s+m,p} \longrightarrow H^{s,p} \tag{2.1.1}$$

provided $0 < s < r$. Furthermore, under these hypotheses,

$$p(x,D) : C^{s+m}_* \longrightarrow C^s_*. \tag{2.1.2}$$

We will present Bourdaud's proof below. First we record some implications. Note that any $p(x,\xi) \in S^m_{1,1}$ satisfies the hypotheses for all $r > 0$. Since operators in $OPS^m_{1,\delta}$ possess good multiplicative properties for $\delta \in [0,1)$, we have the following:

COROLLARY 2.1.B. *If $p(x,\xi) \in S^m_{1,\delta}$, $0 \le \delta < 1$, we have the mapping properties (2.1.1) and (2.1.2) for all $s \in \mathbb{R}$.*

It is known that elements of $OPS^0_{1,1}$ need not be bounded on L^p, even for $p = 2$, but by duality and interpolation we have the following.

COROLLARY 2.1.C. *If $p(x,D)$ and $p(x,D)^*$ belong to $OPS^m_{1,1}$, then (2.1.1) holds for all $s \in \mathbb{R}$.*

The main thesis of Bourdaud [BG] is that $OPS^m_{1,1} \cap (OPS^m_{1,1})^*$ possesses good algebraic properties as well as good mapping properties. See also Hörmander [H2],[H4]; he characterizes these operators by the behavior of their symbols.

The following result will be of great use to us.

PROPOSITION 2.1.D. *If $p \in (1,\infty)$ and $p(x,\xi) \in C^s S^m_{1,0}$, then*

$$\begin{aligned} p(x,D) &: H^{r+m,p} \longrightarrow H^{r,p} \\ p(x,D) &: C^{r+m}_* \longrightarrow C^r_*, \end{aligned} \tag{2.1.3}$$

privided $-s < r < s$.

PROOF: It suffices to take $m = 0$. The result follows from (2.1.1) if $0 < r < s$, so it remains to consider $r \in (-s, 0]$. For this, we make the decomposition (1.3.5), i.e.,

$$p(x,\xi) = p^\#(x,\xi) + p^b(x,\xi),\ p^\#(x,\xi) \in S^0_{1,\delta},\ p^b(x,\xi) \in C^s S^{-s\delta}_{1,\delta}, \tag{2.1.4}$$

with $\delta \in (0,1)$. Now Corollary 2.1.B applies to $p^\#(x,D)$. Meanwhile, applying (2.1.1) to $p^b(x,D)$ with $m = -s\delta$ gives

$$p^b(x,D) : H^{\sigma - s\delta,p} \longrightarrow H^{\sigma,p}(: C^{\sigma - s\delta}_* \longrightarrow C^\sigma_*), \quad 0 < \sigma < s.$$

Picking δ close to 1 then yields (2.1.3) (with $m = 0$) for $-s < r \le 0$.

The following extension will also be useful.

PROPOSITION 2.1.E. *If $p(x,\xi) \in C^s S^0_{1,\delta}$, with $s > 0$, $\delta \in (0,1)$, then, for $1 < p < \infty$,*

$$p(x,D) : H^{r,p} \longrightarrow H^{r,p}(: C^r_* \longrightarrow C^r_*) \text{ for } -(1-\delta)s < r < s. \tag{2.1.5}$$

PROOF: Use the decomposition $p = p^\# + p^b$ having the property (1.3.21), with $\delta < \gamma < 1$, $m = 0$. Applying Proposition 2.1.D to $p^b(x,D)$, and letting $\gamma \to 1$, we obtain (2.1.5).

We prepare to prove Theorem 2.1.A. Following [BG], and also [Ma], we make use of the following results from Littlewood-Paley theory, whose proofs can be found in §A.1.

LEMMA 2.1.F. *Let $f_k \in \mathcal{S}'(\mathbb{R}^n)$ be such that, for some $A > 0$,*

$$supp\ \hat{f}_k \subset \{\xi : A \cdot 2^{k-1} \le |\xi| \le A \cdot 2^{k+1}\}, \quad k \ge 1. \tag{2.1.6}$$

Say $\hat{f}_0$ has compact support. Then, for $p \in (1,\infty)$, $s \in \mathbb{R}$, we have

$$\Big\|\sum_{k=0}^{\infty} f_k\Big\|_{H^{s,p}} \sim \Big\|\Big\{\sum_{k=0}^{\infty} 4^{ks}|f_k|^2\Big\}^{\frac{1}{2}}\Big\|_{L^p}. \tag{2.1.7}$$

LEMMA 2.1.G. *Let $f_k \in \mathcal{S}'(\mathbb{R}^n)$ be such that*

$$supp\ \hat{f}_k \subset \{\xi : |\xi| \le A \cdot 2^{k+1}\}, \quad k \ge 0. \tag{2.1.8}$$

Then, for $p \in (1,\infty)$, $s > 0$, we have

$$\Big\|\sum_{k=0}^{\infty} f_k\Big\|_{H^{s,p}} \le C\Big\|\Big\{\sum_{k=0}^{\infty} 4^{ks}|f_k|^2\Big\}^{\frac{1}{2}}\Big\|_{L^p}. \tag{2.1.9}$$

The next ingredient is a symbol decomposition to replace (1.1.16). This is necessarily more complicated for symbols in $C^r_* S^m_{1,1}$ than for $X^s S^m_{cl}$. We begin with the S^0_1-partition of unity (1.3.1), and with

$$p(x,\xi) = \sum_{j=0}^{\infty} p(x,\xi)\psi_j(\xi) = \sum_{j=0}^{\infty} p_j(x,\xi). \tag{2.1.10}$$

Now, let us take a basis of $L^2(\frac{1}{2} < |\xi| < 2)$ of the form

$$|\xi|^{\frac{1}{2}(1-n)} e^{4\pi i k|\xi|/3} \omega_l(\xi/|\xi|) = \beta_{kl}(\xi),$$

and write (for $j \ge 1$)

$$p_j(x,\xi) = \sum_{k,l} p_{jkl}(x)\beta_{kl}(2^{-j}\xi)\psi_j^{\#}(\xi), \tag{2.1.11}$$

where $\psi_1^{\#}(\xi)$ has support on $\frac{1}{2} < |\xi| < 2$ and is 1 on supp ψ_1, $\psi_j^{\#}(\xi) = \psi_1^{\#}(2^{-j+1}\xi)$, with an analogous decomposition for $p_0(\xi)$. Inserting these decompositions into (2.1.10) and summing over j, we obtain $p(x,\xi)$ as a sum of a rapidly decreasing sequence of elementary symbols. By definition, an elementary symbol in $C^r_* S^0_{1,\delta}$ is of the form

$$q(x,\xi) = \sum_{k=0}^{\infty} Q_k(x)\varphi_k(\xi), \tag{2.1.12}$$

where φ_k is supported on $\langle\xi\rangle \sim 2^k$ and bounded in S^0_1, in fact $\varphi_k(\xi) = \varphi_1(2^{-k+1}\xi)$, for $k \geq 2$, and $Q_k(x)$ satisfies

$$|Q_k(x)| \leq C, \ \|Q_k\|_{C^r} \leq C \cdot 2^{kr\delta}. \tag{2.1.13}$$

For the purpose of proving Theorem 2.1.A, we take $\delta = 1$. It suffices to estimate the $H^{r,p}$-operator norm of $q(x,D)$ when $q(x,\xi)$ is such an elementary symbol.

Set $Q_{kj}(x) = \psi_j(D)Q_k(x)$, with $\{\psi_j\}$ the S^0_1 partition of unity described above. Set

$$\begin{aligned} q(x,\xi) &= \sum_k \{\sum_{j=0}^{k-4} Q_{kj}(x) + \sum_{j=k-3}^{k+3} Q_{kj}(x) + \sum_{j=k+4}^{\infty} Q_{kj}(x)\}\varphi_k(\xi) \\ &= q_1(x,\xi) + q_2(x,\xi) + q_3(x,\xi). \end{aligned} \tag{2.1.14}$$

To estimate $q(x,D)f$, let $f_k = \varphi_k(D)f$. By Lemma 2.1.F, since $\langle\xi\rangle \sim 2^j$ on the spectrum of Q_{kj},

$$\begin{aligned} \|q_1(x,D)f\|_{H^{s,p}} &\leq C\Big\|\{\sum_{k=4}^{\infty} 4^{ks}|\sum_{j=0}^{k-4} Q_{kj}f_k|^2\}^{\frac{1}{2}}\Big\|_{L^p} \\ &\leq C\Big\|\{\sum_{k=4}^{\infty} 4^{ks}|f_k|^2\}^{\frac{1}{2}}\Big\|_{L^p} \\ &\leq C\|f\|_{H^{s,p}}, \end{aligned} \tag{2.1.15}$$

for all $s \in \mathbb{R}$.

To estimate $q_2(x,D)f$, note that $\|Q_{kj}\|_{L^\infty} \leq C \cdot 2^{-jr+kr}$. Then Lemma 2.1.G implies

$$\|q_2(x,D)f\|_{H^{s,p}} \leq C\Big\|\{\sum_{k=0}^{\infty} 4^{ks}|f_k|^2\}^{\frac{1}{2}}\Big\|_{L^p} \leq C\|f\|_{H^{s,p}}, \tag{2.1.16}$$

for $s > 0$.

To estimate $q_3(x,D)f$, we again apply Lemma 2.1.F, to obtain

$$\begin{aligned} \|q_3(x,D)f\|_{H^{s,p}} &\leq C\Big\|\{\sum_{j=4}^{\infty} 4^{js}|\sum_{k=0}^{j-4} Q_{kj}f_k|^2\}^{\frac{1}{2}}\Big\|_{L^p} \\ &\leq C\Big\|\Big\{\sum_{j=4}^{\infty} 4^{j(s-r)}\Big(\sum_{k=0}^{j-4} 2^{kr}|f_k|\Big)^2\Big\}^{\frac{1}{2}}\Big\|_{L^p}. \end{aligned} \tag{2.1.17}$$

Now, if we set $g_j = \sum_{k=0}^{j-4} 2^{(k-j)r}|f_k|$ and then set $G_j = 2^{js}g_j$ and $F_j = 2^{js}|f_j|$, we see that $G_j = \sum_{k=0}^{j-4} 2^{(k-j)(r-s)}F_j$. As long as $r > s$, Young's inequality yields $\|(G_j)\|_{\ell^2} \le C\|(F_j)\|_{\ell^2}$, so the last line in (2.1.17) is bounded by

$$C\|\{\sum_{j=0}^{\infty} 4^{js}|f_j|^2\}^{\frac{1}{2}}\|_{L^p} \le C\|f\|_{H^{s,p}}.$$

This proves (2.1.1).

The proof of (2.1.2) is similar. We replace (2.1.7) by

$$\|f\|_{C^r_*} \sim \sup_{k\ge 0} 2^{kr}\|\psi_k(D)f\|_{L^\infty} \tag{2.1.18}$$

We also need an analogue of Lemma 2.1.G:

LEMMA 2.1.H. *If $f_k \in \mathcal{S}'(\mathbb{R}^n)$ and supp $\hat{f}_k \subset \{\xi : |\xi| \le A\cdot 2^{k+1}\}$, then, for $r > 0$,*

$$\|\sum_{k=0}^{\infty} f_k\|_{C^r_*} \le C \sup_{k\ge 0} 2^{kr}\|f_k\|_{L^\infty}. \tag{2.1.19}$$

PROOF: As $\psi_j(D)(\sum_{k=0}^{\infty} f_k)$ is a finite sum (with a bounded number of terms) of $\psi_j(D)f_k$, over $j - N \le k \le j + N$, this follows easily from (2.1.18).

Now, to prove (2.1.2), as before it suffices to consider elementary symbols, of the form (2.1.12) - (2.1.13), and we use again the decomposition $q(x,\xi) = q_1 + q_2 + q_3$ of (2.1.14). Thus it remains to obtain analogues of the estimates (2.1.15) - (2.1.17).

Parallel to (2.1.15), using the fact that $\sum_{j=0}^{k-4} Q_{kj}(x)f_k$ has spectrum in $\langle\xi\rangle \sim 2^k$, and $\|Q_k\|_{L^\infty} \le C$, we obtain

$$\begin{aligned} \|q_1(x,D)f\|_{C^s_*} &\le C \sup_{k\ge 0} 2^{ks}\|\sum_{j=0}^{k-4} Q_{kj}f_k\|_{L^\infty} \\ &\le C \sup_{k\ge 0} 2^{ks}\|f_k\|_{L^\infty} \\ &\le C\|f\|_{C^s_*}, \end{aligned} \tag{2.1.20}$$

for all $s \in \mathbb{R}$. Parallel to (2.1.16), using $\|Q_{kj}\|_{L^\infty} \le C \cdot 2^{-jr+kr}$ and Lemma 2.1.H, we have

$$
\begin{aligned}
\|q_2(x,D)f\|_{C^s_*} &\le \Big\|\sum_{k=0}^{\infty} g_k\Big\|_{C^s_*} \\
&\le C \sup_{k\ge 0} 2^{ks}\|g_k\|_{L^\infty} \\
&\le C \sup_{k\ge 0} 2^{ks}\|f_k\|_{L^\infty} \le C\|f\|_{C^s_*},
\end{aligned}
\tag{2.1.21}
$$

for all $s > 0$, where the sum of 7 terms

$$
g_k = \sum_{j=k-3}^{k+3} Q_{kj}(x) f_k
$$

has spectrum contained in $|\xi| \le C \cdot 2^k$, and $\|g_k\|_{L^\infty} \le C$.

Finally, parallel to (2.1.17), as $\sum_{k=0}^{j-4} Q_{kj} f_k$ has spectrum in $\langle\xi\rangle \sim 2^j$, we have

$$
\begin{aligned}
\|q_3(x,D)f\|_{C^s_*} &\le C \sup_{j\ge 0} 2^{js}\Big\|\sum_{k=0}^{j-4} Q_{kj} f_k\Big\|_{L^\infty} \\
&\le C \sup_{j\ge 0} 2^{j(s-r)} \sum_{k=0}^{j-4} 2^{kr}\|f_k\|_{L^\infty}.
\end{aligned}
\tag{2.1.22}
$$

If we bound this last sum by

$$
\Big(\sum_{k=0}^{j-4} 2^{k(r-s)}\Big) \sup_k 2^{ks}\|f_k\|_{L^\infty},
\tag{2.1.23}
$$

then

$$
\|q_3(x,D)f\|_{C^s_*} \le C\Big(\sup_{j\ge 0} 2^{j(s-r)} \sum_{k=0}^{j-4} 2^{k(r-s)}\Big)\|f\|_{C^s_*}
\tag{2.1.24}
$$

and the factor in brackets is finite as long as $s < r$. The proof of Theorem 2.1.A is complete.

Things barely blow up in (2.1.24) when $s = r$. The following result is of some use.

PROPOSITION 2.1.I. *If* $p(x,\xi) \in C^r S^0_{1,1}$, *then*

$$p(x,D) : C^{r+\epsilon}_* \longrightarrow C^r_* \text{ for all } \epsilon > 0. \tag{2.1.25}$$

PROOF: We follow the proof of (2.1.2). The estimates (2.1.20) and (2.1.21) continue to work; (2.1.22) yields

$$\begin{aligned} \|q_3(x,D)f\|_{C^r_*} &\leq C \sup_{j\geq 0} \sum_{k=0}^{j-4} 2^{kr} \|f_k\|_{L^\infty} \\ &= C \sum_{k=0}^{\infty} 2^{kr} \|f_k\|_{L^\infty} \\ &\leq C \sum_{k=0}^{\infty} 2^{kr} \cdot 2^{-kr-k\epsilon} \|f\|_{C^{r+\epsilon}_*} \end{aligned} \tag{2.1.26}$$

which proves (2.1.25).

Extensions of Bourdaud's results have been obtained by Marschall [Ma]. Some of them are summarized as follows:

PROPOSITION 2.1.J. *Suppose* $1 \leq q \leq \infty$, $s > \frac{n}{q}$, $1 < p < \infty$, $m \in \mathbb{R}$. *Then*

$$p(x,\xi) \in H^{s,q} S^m_{1,0} \Longrightarrow p(x,D) : H^{r+m,p} \longrightarrow H^{r,p} \tag{2.1.27}$$

provided

$$n\Big(\frac{1}{p} + \frac{1}{q} - 1\Big)^+ - s < r \leq s - n\Big(\frac{1}{q} - \frac{1}{p}\Big)^+. \tag{2.1.28}$$

Note that, when $q = p$, the condition (2.1.28) becomes

$$n\Big(\frac{2}{p} - 1\Big)^+ - s < r \leq s. \tag{2.1.29}$$

We refer to [Ma] for a proof of this and other results. Marschall also defines other symbol classes, of the form $X^s S^m_{1,\delta}$, analogous to the definition (1.3.16) - (1.3.17), for $X^s = H^{s,p}, \delta \in [0,1]$, and has further results on these symbols. We also mention related results of Nagase [N] and Kumano-go and Nagase [KN].

The case $q = p = 2$ of Proposition 2.1.J had been obtained by Beals-Reed [BR]. We include their short proof here. As in [BR], we make use of the following:

LEMMA 2.1.K. *Suppose that*

$$\sup_{\xi} \int |g(\eta,\xi)|^2 d\eta = A^2 < \infty \tag{2.1.30}$$

and

$$\sup_{\eta} \int |G(\eta,\xi)|^2 d\xi = B^2 < \infty. \tag{2.1.31}$$

Then

$$Tf(\eta) = \int G(\eta,\xi) g(\eta-\xi,\eta) f(\xi) d\xi \tag{2.1.32}$$

satisfies

$$\|Tf\|_{L^2} \le A \cdot B \, \|f\|_{L^2}. \tag{2.1.33}$$

PROOF: Simple consequence of the Schwartz inequality.

We now use this to prove Proposition 2.1.J when $q = p = 2$ and $r = s$. It suffices to consider the case $m = 0$. We will also assume $p(x,\xi)$ has compact support in x. Then

$$(p(x,D)u)\hat{}\,(\eta) = \int \hat{p}(\eta-\xi,\xi)\hat{u}(\xi) d\xi, \tag{2.1.34}$$

where $\hat{p}$ denotes the partial Fourier transform with respect to x. The hypothesis $p \in H^s S^0_{1,0}$ implies

$$\hat{p}(\zeta,\xi) = g(\zeta,\xi)\langle\zeta\rangle^{-s}, \text{ with (2.1.30) holding for } g. \tag{2.1.35}$$

If $f = \langle\xi\rangle^s \hat{u}$, then $u \in H^s$ implies $f \in L^2$, and we have

$$\langle\eta\rangle^s (p(x,D)u)\hat{}\,(\eta) = \int G(\eta,\xi) g(\eta-\xi,\xi) f(\xi) d\xi \tag{2.1.36}$$

where

$$G(\eta,\xi) = \langle\eta\rangle^s / [\langle\eta-\xi\rangle^s \langle\xi\rangle^s]. \tag{2.1.37}$$

One can show that G satisfies (2.1.31), granted that $s > \frac{n}{2}$, so Lemma 2.1.K gives a bound for the L^2-norm of (2.1.36), hence a bound for $\|p(x,D)u\|_{H^s}$, as desired.

§2.2. Further elliptic regularity theorems

We return to the setting of §1.2, first considering an elliptic differential operator of order m

$$A(x,D) = \sum_{|\alpha|\le m} a_\alpha(x)\, D^\alpha, \tag{2.2.1}$$

whose coefficients have limited regularity. We will eschew the generality of §1.2 and concentrate on $X^s = C^s$. Thus $A(x,\xi) \in C^s S^m_{cl}$ is elliptic. Pick $\delta \in (0,1)$ and write

$$A(x,\xi) = A^\#(x,\xi) + A^b(x,\xi) \tag{2.2.2}$$

with

$$A^\#(x,\xi) \in S^m_{1,\delta}, \quad A^b(x,\xi) \in C^s S^{m-\delta s}_{1,\delta}. \tag{2.2.3}$$

Consequently, by Proposition 2.1.E,

$$A^b(x,D) : C_*^{m+r-\delta s} \longrightarrow C_*^r, \quad -(1-\delta)s < r < s. \tag{2.2.4}$$

Now let $p(x,D) \in OPS^{-m}_{1,\delta}$ be a parametrix for $A^\#(x,D)$, which is elliptic. Hence, mod C^∞,

$$p(x,D)A(x,D)u = u + p(x,D)A^b(x,D)u, \tag{2.2.5}$$

so if

$$A(x,D)u = f, \tag{2.2.6}$$

then, mod C^∞,

$$u = p(x,D)f - p(x,D)A^b(x,D)u. \tag{2.2.7}$$

In view of (2.2.5), we see that, when (2.2.7) is satisfied,

$$u \in C_*^{m+r-\delta s},\ f \in C_*^r \Longrightarrow u \in C_*^{m+r}. \tag{2.2.8}$$

We then have the following.

PROPOSITION 2.2.A. *Let $A(x,\xi) \in C^s S^m_{cl}$ be elliptic and suppose u solves (2.2.6). Assuming*

$$s > 0, \quad 0 < \delta < 1, \quad \text{and } -(1-\delta)s < r < s, \tag{2.2.9}$$

we have

$$u \in C^{m+r-\delta s}, \ f \in C^r_* \Longrightarrow u \in C^{m+r}_*. \tag{2.2.10}$$

Note that, for $|\alpha| = m$, $D^\alpha u \in C^{r-\delta s}_*$, and $r - \delta s$ could be negative. However, $a_\alpha(x) D^\alpha u$ will still be well defined for $a_\alpha \in C^s$. Indeed, if (2.1.3) is applied to the special case of a multiplication operator, we have

$$a \in C^s, \ u \in C^\sigma_* \Longrightarrow au \in C^\sigma_* \text{ for } -s < \sigma < s. \tag{2.2.11}$$

Note that the range of r in (2.2.9) can be rewritten as $-s < r - \delta s < (1-\delta)s$. If we set $r - \delta s = -s + \epsilon$, this means $0 < \epsilon < (2-\delta)s$, so we can rewrite (2.2.10) as

$$u \in C^{m-s+\epsilon}, \ f \in C^r_* \Longrightarrow u \in C^{m+r}_*, \text{ provided } \epsilon > 0, \ r < s, \tag{2.2.12}$$

as long as the relation $r = -(1-\delta)s + \epsilon$ holds. Letting δ range over $(0,1)$, we see that this will hold for any $r \in (-s+\epsilon, \epsilon)$. However, if $r \in [\epsilon, s)$, we can first obtain from the hypothesis (2.2.12) that $u \in C^{m+\rho}_*$, for any $\rho < \epsilon$. This improves the a priori regularity of u by almost s units. This argument can be iterated repeatedly, to yield:

THEOREM 2.2.B. *If $A(x,\xi) \in C^s S^m_{cl}$ is elliptic and u solves (2.2.6), and if $s > 0$ and $\epsilon > 0$, then*

$$u \in C^{m-s+\epsilon}, \ f \in C^r_* \Longrightarrow u \in C^{m+r}_* \text{ for } -s < r < s. \tag{2.2.13}$$

We can sharpen this up to obtain the following Schauder regularity result.

THEOREM 2.2.C. *Under the hypotheses above,*

$$u \in C^{m-s+\epsilon}, \ f \in C^s_* \Longrightarrow u \in C^{m+s}_*. \tag{2.2.14}$$

PROOF: Applying (2.2.13), we can assume $u \in C^{m+r}_*$ with $s - r > 0$ arbitrarily small. Now if we invoke Proposition 2.1.I, we can supplement (2.2.4) with

$$A^b(x,D) : C^{m+s-\delta s+\epsilon}_* \longrightarrow C^s_*, \quad \epsilon > 0.$$

If $\delta > 0$, and if $\epsilon > 0$ is picked small enough, we can write $m + s - \delta s + \epsilon = m + r$ with $r < s$, so, under the hypotheses of (2.2.14), the right side of (2.2.7) belongs to C^{m+s}_*, proving the theorem. We note that a similar argument also produces the regularity result:

$$u \in H^{m-s+\epsilon,p}, \ f \in C^s_* \Longrightarrow u \in C^{m+s}_*. \tag{2.2.15}$$

As in §1.2, we apply this to solutions to the quasilinear elliptic PDE

$$\sum_{|\alpha|\le m} a_\alpha(x, D^{m-1}u)\, D^\alpha u = f. \tag{2.2.16}$$

As long as $u \in C^{m-1+s}$, $a_\alpha(x, D^{m-1}u) \in C^s$. If also $u \in C^{m-s+\epsilon}$, we obtain (2.2.13) and (2.2.14). If $r > s$, using the conclusion $u \in C_*^{m+s}$, we obtain $a_\alpha(x, D^{m-1}u) \in C^{s+1}$, so we can reapply (2.2.13)-(2.2.14) for further regularity, obtaining the following.

THEOREM 2.2.D. *If u solves the quasilinear elliptic PDE* (2.2.16), *then*

$$u \in C^{m-1+s} \cap C^{m-s+\epsilon},\ f \in C_*^r \Longrightarrow u \in C_*^{m+r} \tag{2.2.17}$$

provided $s > 0$, $\epsilon > 0$, and $-s < r$. Thus

$$u \in C^{m-1+s},\ f \in C_*^r \Longrightarrow u \in C_*^{m+r} \tag{2.2.18}$$

provided

$$s > \frac{1}{2},\ r > s - 1. \tag{2.2.19}$$

We can sharpen up Theorem 2.2.D a bit as follows. Replace the hypothesis in (2.2.17) by

$$u \in C^{m-1+s} \cap H^{m-1+\sigma,p}, \tag{2.2.20}$$

with $p \in (1, \infty)$. Parallel to (2.2.11), we have

$$a \in C^s,\ u \in H^{\sigma,p} \Longrightarrow au \in H^{\sigma,p},\ \text{for } -s < \sigma < s, \tag{2.2.21}$$

as a consequence of (2.1.3), so we see that the left side of (2.2.16) is well defined provided $s + \sigma > 1$. We have (2.2.7), with

$$A^b(x, D) : H^{m+r-\delta s,p} \longrightarrow H^{r,p} \text{ for } -(1-\delta)s < r < s, \tag{2.2.22}$$

parallel to (2.2.4). Thus, if (2.2.20) holds, we obtain

$$p(x, D)A^b(x, D)u \in H^{m-1+\sigma+\delta s,p} \tag{2.2.23}$$

provided $-(1-\delta)s < \delta s - 1 + \sigma < s$, i.e., provided

$$s + \sigma > 1 \text{ and } -1 + \sigma + \delta s < s. \tag{2.2.24}$$

Thus, if $f \in H^{\rho,p}$ with $\rho > \sigma - 1$, we manage to improve the regularity of u over the hypothesized (2.2.20). One way to record this gain is to use the Sobolev imbedding theorem:

$$H^{m-1+\sigma+\delta s,p} \subset H^{m-1+\sigma,p_1}, \quad p_1 = \frac{pn}{n-\delta s} > p\Big(1 + \frac{\delta sp}{n}\Big). \tag{2.2.25}$$

If we assume $f \in C_*^r$ with $r > \sigma - 1$, we can iterate this argument sufficiently often to obtain $u \in C^{m-1+\sigma-\epsilon}$, for arbitrary $\epsilon > 0$. Now we can arrange $s + \sigma > 1 + \epsilon$, so we are now in a position to apply Theorem 2.2.D. This proves:

THEOREM 2.2.E. *If u solves the quasilinear elliptic PDE (2.2.16), then*

$$u \in C^{m-1+s} \cap H^{m-1+\sigma,p}, \ f \in C_*^r \Longrightarrow u \in C_*^{m+r} \tag{2.2.26}$$

provided $1 < p < \infty$ and

$$s > 0, \ s + \sigma > 1, \ r > \sigma - 1. \tag{2.2.27}$$

Note that if $u \in H^{m,p}$ for some $p > n$, then $u \in C^{m-1+s}$ for $s = 1 - \frac{n}{p} > 0$, and then (2.2.26) applies, with $\sigma = 1$, or even with $\sigma = \frac{n}{p} + \epsilon$.

We compare Theorem 2.2.E with material in Chapter 9 of Gilbarg and Trudinger [GT], treating the case of a scalar elliptic PDE of order 2. In that case, if u is a solution to (2.2.16), Theorem 9.13 - Lemma 9.16 of [GT] imply the following. Assume $u \in C^1$ and $u \in H^{2,p}$ for some $p \in (1, \infty)$. Then, given any $q \in (p, \infty)$, if $f \in L^q$, then $u \in H^{2,q}$. If $q > n$, we can apply the observation above, to conclude:

THEOREM 2.2.F. *If $m = 2$ and (2.2.16) is scalar, then, given $p \in (1, \infty)$, $q > n$,*

$$u \in C^1 \cap H^{2,p}, \ f \in L^q \cap C_*^r \Longrightarrow u \in C_*^{2+r} \ \text{if}\ r > -1 + \frac{n}{q}. \tag{2.2.28}$$

We note parenthetically that $L^q \subset C_*^\rho$ for $\rho < -\frac{n}{q}$.

We also record the following improvement of Theorem 1.2.D, regarding the regularity of solutions to a completely nonlinear elliptic system

$$F(x, D^m u) = f. \tag{2.2.29}$$

We could apply Theorem 2.2.B - Theorem 2.2.C to the equation for $u_j = \partial u / \partial x_j$:

$$\sum_{|\alpha| \le m} (\partial F / \partial \zeta_\alpha)(x, D^m u) D^\alpha u_j = -F_{x_j}(x, D^m u) + \partial f / \partial x_j = f_j. \tag{2.2.30}$$

Suppose $u \in C^{m+s}$, $s > 0$, so the coefficients $a_\alpha(x) = (\partial F/\partial \zeta_\alpha)(x, D^m u) \in C^s$. If $f \in C^r_*$, then $f_j \in C^s \cup C^{r-1}_*$. We can apply Theorem 2.2.B - Theorem 2.2.C to u_j provided $u \in C^{m+1-s+\epsilon}$, to conclude that $u \in C^{m+s+1}_* \cup C^{m+r}_*$. This implication can be iterated as long as $s + 1 < r$, until we obtain $u \in C^{m+r}_*$.

This argument has the drawback of requiring too much regularity of u, namely that $u \in C^{m+1-s+\epsilon}$ as well as $u \in C^{m+s}$. We can fix this up by considering difference quotients rather than derivatives $\partial_j u$. Thus, for $y \in \mathbb{R}^n$, $|y|$ small, set

$$v_y(x) = |y|^{-1}\big[u(x+y) - u(x)\big];$$

v_y satisfies the PDE

$$\sum_{|\alpha| \le m} \Phi_{\alpha y}(x) D^\alpha v_y(x) = G_y(x, D^m u) \tag{2.2.31}$$

where

$$\Phi_{\alpha y}(x) = \int_0^1 (\partial F/\partial \zeta_\alpha)\big(x, tD^m u(x) + (1-t)D^m u(x+y)\big)\,dt \tag{2.2.32}$$

and G_y is an appropriate analogue of the right side of (2.2.30). Thus $\Phi_{\alpha y}$ is in C^s, uniformly as $|y| \to 0$, if $u \in C^{m+s}$, while this hypothesis also gives a uniform bound on the C^{m-1+s}-norm of v_y. Now, for each y, Theorems 2.2.B and 2.2.C apply to v_y, and one can get an *estimate* on $\|v_y\|_{C^{m+\rho}}$, $\rho =$ min $(s, r-1)$, *uniform* as $|y| \to 0$. Therefore we have the following.

THEOREM 2.2.G. *If u solves the elliptic PDE* (2.2.29), *then*

$$u \in C^{m+s}, \ f \in C^r_* \Longrightarrow u \in C^{m+r}_* \tag{2.2.33}$$

provided

$$0 < s < r. \tag{2.2.34}$$

Another proof of this result will be given in §3.3; see Theorem 3.3.C.

We briefly discuss results for PDE in divergence form which hold for less regular u than required in Theorem 2.2.B - Theorem 2.2.E. We restrict attention to the case $m = 2$. Thus consider

$$\sum_{j,k} \partial_j a_{jk}(x) \partial_k u = f, \tag{2.2.35}$$

which we assume to be elliptic. We can write this as

$$\sum_j \partial_j A_j(x,D)u = f, \tag{2.2.36}$$

where

$$a_{jk}(x) \in C^r \Longrightarrow A_j(x,\xi) \in C^r S^1_{1,0}. \tag{2.2.37}$$

Using the decomposition

$$\begin{gathered} A_j(x,\xi) = A_j^{\#}(x,\xi) + A_j^b(x,\xi), \\ A_j^{\#}(x,\xi) \in S^1_{1,\delta},\ A_j^b(x,\xi) \in C^r S^{1-\delta r}_{1,\delta}, \end{gathered} \tag{2.2.38}$$

with $0 < \delta < 1$, we have

$$\sum \partial_j A_j^{\#}(x,D) = B(x,D) \in OPS^2_{1,\delta} \text{ elliptic }. \tag{2.2.39}$$

Let $p(x,D) \in OPS^{-2}_{1,\delta}$ be a parametrix for $B(x,D)$. Then (2.2.36) implies

$$u = p(x,D)f - \sum p(x,D)\partial_j A_j^b(x,D)u \bmod C^\infty. \tag{2.2.40}$$

Now, for $q \in (1,\infty)$, Proposition 2.1.E implies

$$A_j^b(x,D) : H^{\sigma,q} \Longrightarrow H^{\sigma-1+\delta r,q} \tag{2.2.41}$$

provided $-(1-\delta)r < \sigma - 1 + \delta r < r$, i.e., provided

$$1 - r < \sigma < 1 + (1-\delta)r. \tag{2.2.42}$$

With these results in mind, we can establish the following, which gives an affirmative answer to a question of Jeff Cheeger.

THEOREM 2.2.H. *Suppose the PDE (2.2.35) is elliptic, with*

$$u \in H^{\sigma,q},\ f \in H^{-1,p},\ a_{jk} \in C^r, \tag{2.2.43}$$

where

$$1 < q < p < \infty,\ r > 0 \text{ and } \sigma > 1 - r. \tag{2.2.44}$$

Then

$$u \in H^{1,p}. \tag{2.2.45}$$

PROOF: Looking at (2.2.40), we have $p(x,D)f \in H^{1,p}$. If $\sigma < 1$, (2.2.40) yields $u \in H^{\tilde{\sigma},q}$ with $\tilde{\sigma} = \min(\sigma + \delta r, 1)$. We can iterate this to get $u \in H^{1,q}$. From there, we can apply (2.2.41) with $\sigma = 1$, to obtain $A_j^b(x,D)u \in H^{\delta r,q}$, and hence

$$p(x,D)\partial_j A_j^b(x,D)u \in H^{1+\delta r,q} \subset H^{1,\tilde{q}} \tag{2.2.46}$$

where

$$\tilde{q} = \frac{nq}{n - \delta r q} > q\Big(1 + \frac{\delta r q}{n}\Big). \tag{2.2.47}$$

Hence $u \in H^{1,q_1}$ with $q_1 = \min(p, \tilde{q})$. Iterating this argument a finite number of times we obtain the desired property (2.2.45).

One use for this is in the study of the Ricci tensor. If one uses local *harmonic* coordinates on a Riemannian manifold, then the metric tensor and Ricci tensor are related by

$$-\frac{1}{2}\sum_{j,k} g^{jk}\partial_j\partial_k g_{\ell m} + Q_{\ell m}(g, Dg) = R_{\ell m}, \tag{2.2.48}$$

where $Q_{\ell m}(g,\zeta)$ is a certain quadratic form in ζ, with coefficients smooth in g. We can rewrite this as

$$-\frac{1}{2}\sum_{j,k} \partial_j g^{jk}\partial_k g_{\ell m} + Q'_{\ell m}(g, Dg) = R_{\ell m}, \tag{2.2.49}$$

with a slightly different $Q'_{\ell m}$ of the same nature. The goal is to presume a priori some weak estimates on g_{jk}, and, given certain regularity of $R_{\ell m}$, deduce better estimates on the metric tensor, in this coordinate system. Thus it is useful to supplement Theorem 2.2.H with the following.

We consider a PDE of the form

$$\sum \partial_j a_{jk}(x,u)\partial_k u + B(x,u,Du) = f, \tag{2.2.50}$$

assumed to be elliptic on a region in $\mathbb{R}^n$. Assume $B(x,u,Du)$ is a quadratic form in Du, with coefficients smooth in x, u, and $a_{jk}(x,u)$ smooth in its arguments.

PROPOSITION 2.2.I. *Let the elliptic PDE* (2.2.50) *be solved by*

$$u \in H^{1,q}, \quad q > n. \tag{2.2.51}$$

Let $q < p < \infty$ and assume

$$f \in H^{-1,p}. \tag{2.2.52}$$

Then

$$u \in H^{1,p}. \tag{2.2.53}$$

PROOF: Note that (2.2.51) implies that $u \in C^r$ for some $r > 0$, and hence

$$B(x, u, Du) \in L^{q/2}. \tag{2.2.54}$$

Rewrite (2.2.50) as

$$\sum \partial_j a_{jk}(x) \partial_k u = g = f - B, \tag{2.2.55}$$

with $a_{jk}(x) = a_{jk}(x, u) \in C^r$. From (2.2.52), (2.2.54), and the Sobolev imbedding theorem,

$$L^{q/2} \subset H^{-1,s}, \quad s = \frac{\frac{1}{2}q}{1 - \frac{q}{2n}},$$

we deduce that $g \in H^{-1,\tilde{q}}$, $\tilde{q} = \min(s, p)$. Then Theorem 2.2.H implies $u \in H^{1,\tilde{q}}$. This is an improvement over the hypothesis (2.2.51) if $q > n$, since

$$q = n + a \Longrightarrow 1 - \frac{q}{2n} = \frac{1}{2}\left(1 - \frac{a}{n}\right) \Longrightarrow s > q\left(1 + \frac{a}{n}\right).$$

Iterating this argument gives the conclusion (2.2.53).

This gives results on (2.2.49) complementary to the result of DeTurk and Kazdan [DK] that, if g is C^2 and $R_{\ell m}$ is $C^{k+\alpha}$, in harmonic coordinates, then g is $C^{k+2+\alpha}$. It is clear that further generalizations can be established. For example, if (2.2.50) is elliptic and (2.2.51) holds, then

$$f \in H^{\sigma,p}, \ \sigma \geq -1 \Longrightarrow u \in H^{\sigma+2,p}. \tag{2.2.56}$$

Proposition 2.2.I can be brought to bear on some of the material in [AC].

To end this section, we recall and apply a result established by the DeGeorgi-Nash-Moser theory for the divergence form PDE (2.2.35) in the case when the coefficients $a_{jk}(x)$ are *scalar*. A proof is given in Appendix C.

THEOREM 2.2.J. *Suppose the PDE* (2.2.35) *is elliptic on* Ω *with*

$$a_{jk} \in L^\infty \text{ scalar}, \tag{2.2.57}$$

$$f = g + \sum \partial_j f_j \text{ with } g \in L^{q/2}, \ f_j \in L^q, \ q > n, \tag{2.2.58}$$

and $u \in H^{1,2}$. *Then*

$$u \in C^r \text{ for some } r > 0, \tag{2.2.59}$$

with C^r*-norm on compact sets bounded by* $\|u\|_{L^2}, \|f_j\|_{L^q}, \|g\|_{L^{q/2}}$ *and the ellipticity constants.*

Sometimes this can be used in concert with the last two theorems.

We recall a classical use of Theorem 2.2.J. Given a domain $\Omega \subset \mathbb{R}^n$, with smooth boundary, $\varphi \in H^s(\partial\Omega),\ s \geq \frac{1}{2}$, we seek to minimize

$$I(u) = \int\limits_{\Omega} F(\nabla u)dx \tag{2.2.60}$$

over $V_\varphi^1 = \{u \in H^1(\Omega) : u = \varphi \text{ on } \partial\Omega\}$, assuming that $F(p)$ is smooth in its arguments and satisfies

$$\begin{gathered} C_1|p|^2 - K_1 \leq F(p) \leq C_2|p|^2 + K_2 \\ A_1|\xi|^2 \leq \sum \partial_{p_j}\partial_{p_k}F(p)\xi_j\xi_k \leq A_2|\xi|^2. \end{gathered} \tag{2.2.61}$$

The existence of such a minimum is established in the early part of [Mor]. Such u is a weak solution to the nonlinear PDE

$$\sum_j \partial_{x_j}(\partial_{p_j}F)(\nabla u) = 0. \tag{2.2.62}$$

One next wants to establish higher regularity of u, first on the interior of Ω. Given $y \in \mathbb{R}^n,\ |y|$ small, the difference quotient

$$w_y(x) = |y|^{-1}\left[u(x+y) - u(x)\right]$$

satisfies the divergence form PDE

$$\sum_{j,k} \partial_j a_y^{jk}(x)\partial_k w_y(x) = 0 \tag{2.2.63}$$

with

$$a_y^{jk}(x) = \int_0^1 (\partial_{p_j}\partial_{p_k}F)\big((1-t)\nabla u(x) + t\nabla u(x+y)\big)dt. \tag{2.2.64}$$

Each $w_y \in H^1$, and we have an L^2-bound on w_y over any $\mathcal{O} \subset\subset \Omega$, as $|y| \to 0$. From Theorem 2.2.J there follows a C^r-bound on w_y on any

$\mathcal{O} \subset\subset \Omega$, hence we have $u \in C^{1+r}$ on the interior of Ω. Now $u_\ell = \partial_\ell u$ satisfies

$$\sum_{j,k} \partial_j a^{jk}(x) \partial_k u_\ell = 0, \tag{2.2.65}$$

with

$$a^{jk}(x) = (\partial_{p_j} \partial_{p_k} F)(\nabla u(x)) \in C^r, \tag{2.2.66}$$

so the Schauder theory now kicks in to yield further regularity. Boundary regularity will be examined in §8.3.

We remark on the fact that, in Theorem 2.2.J, even though we require $u \in H^{1,2}$, we can use $\|u\|_{L^2}$ rather than $\|u\|_{H^{1,2}}$ to estimate the C^r-norm. This arises as follows. If $K \subset\subset \Omega$, pick $\psi \in C_0^\infty(\Omega), \psi = 1$ on K, and, for $u \in H^{1,2}_{\text{loc}}(\Omega)$ solving (2.2.35), write

$$\begin{aligned} \sum \int_\Omega \psi(x)^2 a_{jk}(x)(\partial_j u)(\partial_k u) dx \\ = -2 \sum \int_\Omega (\partial_j \psi) \psi a_{jk} (\partial_k u) u dx \\ + \int_\Omega \Big\{ \big[2 \sum (\partial_j \psi) \psi f_j - \psi^2 g\big] u + \sum \psi^2 (\partial_j u) f_j \Big\} dx, \end{aligned} \tag{2.2.67}$$

integrating by parts. In the first term on the right, group together $\psi(\partial_k u)$ and $(\partial_j \psi) u$, and apply Cauchy's inequality. Give a similar treatment to the last term. This leads to an estimate

$$\int_\Omega \psi(x)^2 |\nabla u|^2 dx \le C \int_\Omega \Big[|\nabla \psi|^2 |u|^2 + \sum \psi^2 |f_j|^2 + \psi^2 |g| \cdot |u| \Big] dx, \tag{2.2.68}$$

hence

$$\|\nabla u\|_{L^2(K)} \le C \|u\|^2_{L^2(\Omega)} + C \|g\|^2_{L^2(\Omega)} + C \sum \|f_j\|^2_{L^2(\Omega)} \tag{2.2.69}$$

for a solution to (2.2.35). Here, C depends on the ellipticity constants but not on any regularity of the coefficients a_{jk}.

§2.3. Adjoints

It is useful to understand some things about adjoints of operators with symbols in $C^r S^m_{1,0}$. The results we record here follow simply from the symbol smoothing techniques of §1.3 and mapping properties of §2.1, plus standard results on pseudodifferential operators.

We are particularly interested in $p(x,D)^*$ when $p(x,\xi) \in C^r S^1_{1,0}$. Recall that, in this case,

$$p(x,\xi) = p^{\#}(x,\xi) + p^b(x,\xi), \tag{2.3.1}$$

with

$$p^{\#}(x,\xi) \in S^1_{1,\delta} \tag{2.3.2}$$

and, for $r = \ell + \sigma,\ 0 < \sigma < 1$,

$$D^\beta_x p^{\#}(x,\xi) \in S^1_{1,\delta} \text{ for } |\beta| \le \ell. \tag{2.3.3}$$

Meanwhile,

$$p^b(x,\xi) \in C^r S^{1-r\delta}_{1,\delta}. \tag{2.3.4}$$

It follows from Proposition 2.1.D that

$$p^b(x,D) : H^s \longrightarrow H^s \text{ for } -r < s < r, \text{ if } r\delta \ge 1. \tag{2.3.5}$$

In view of (2.3.3), the standard symbol expansion for $p^{\#}(x,D)^*$ gives

$$p^{\#}(x,D)^* - q^{\#}(x,D) \in OPS^0_{1,\delta} \text{ if } r > 1, \tag{2.3.6}$$

with

$$q^{\#}(x,\xi) = p^{\#}(x,\xi)^*. \tag{2.3.7}$$

Noting (2.3.4) - (2.3.5), we deduce:

PROPOSITION 2.3.A. *Given $p(x,\xi) \in C^r S^1_{1,0}$, $r > 1$, we have*

$$p(x,D)^* - q(x,D) : H^s \longrightarrow H^s \text{ for } -r < s < r \tag{2.3.8}$$

with

$$q(x,\xi) = p(x,\xi)^*. \tag{2.3.9}$$

§2.4. Sharp Gårding inequality

Let $p(x,\xi) \in C^s S^m_{1,0}$ be scalar, with $p(x,\xi) \geq -C_0$. We aim to show that, for a certain range of positive m, $p(x, D)$ is semi-bounded on L^2; this is a sharp Gårding inequality. We will derive it simply by decomposing $p(x,\xi)$ and applying known results for pseudodifferential operators. Recall we can write

$$p(x,\xi) = p^{\#}(x,\xi) + p^{b}(x,\xi) \tag{2.4.1}$$

with

$$p^{\#}(x,\xi) \in S^m_{1,\delta}, \quad p^b(x,\xi) \in C^s S^{m-s\delta}_{1,\delta}, \tag{2.4.2}$$

for any given $\delta \in (0,1)$. Furthermore, $p^b(x, D)$ is bounded on L^2, by Proposition 2.1.E, as long as

$$m - \delta s \leq 0. \tag{2.4.3}$$

If this condition holds, it remains only to consider semiboundedness of $p^{\#}(x, D)$, which belongs to $OPS^m_{1,\delta}$. We may as well apply the best available estimate for this, the Fefferman-Phong inequality [FP], which implies $p^{\#}(x, D)$ is semibounded on L^2 as long as

$$m \leq 2(1-\delta). \tag{2.4.4}$$

Thus, we have semiboundedness of $p(x, D)$ on L^2 as long as, for some $\delta \in (0,1)$, we have $m \leq \min\{\delta s, 2(1-\delta)\}$. Maximizing over $0 < \delta < 1$ gives $\frac{2s}{2+s}$ as the optimal value of m. We have proved:

PROPOSITION 2.4.A. *Let $p(x,\xi) \in C^s S^m_{1,0}$ be scalar, and bounded from below, $p(x,\xi) \geq -C_0$. Then, for all $u \in C_0^\infty$,*

$$Re\ (p(x, D)u, u) \geq -C_1\ \|u\|^2_{L^2}, \tag{2.4.5}$$

provided $s > 0$ and $m \leq \frac{2s}{2+s}$.

In particular, this result applies when $p(x,\xi) \in H^{\sigma,2} S^m_{1,0}$, $\sigma = \frac{1}{2}n + s$. In this case, a strengthened version of Lemma 3.1 of Beals-Reed [BR] is obtained. The proposition above can also be compared with Theorem 7.1 in [H2]. There is demonstrated a semiboundedness result for a class of operators with symbols contained in $C^0 S^{m_0}_{1,0} \cap C^2 S^{m_2+2}_{1,0}$, with $m_0 + m_2 \leq 0$. More precisely, using notation to be defined in (3.3.34), it is assumed that $p(x,\xi) \in \tilde{S}^{m_0}_{1,1}$ and, for $|\beta| = 2$, $D^\beta_x p(x,\xi) \in \tilde{S}^{m_2+2}_{1,1}$, with $m_0 + m_2 \leq 0$.

If $p(x,\xi)$ is a positive semidefinite $L \times L$ matrix, then we cannot appeal to Fefferman-Phong, so instead of (2.4.4) we must require $m \le 1-\delta$. Thus, in this case, we have semiboundedness of (2.4.5) provided $s > 0$ and

$$m \le \frac{s}{1+s}. \tag{2.4.6}$$

We will also have occasion to use the folowing ordinary Gårding inequality, valid for $L \times L$ systems.

PROPOSITION 2.4.B. *Let $p(x,\xi) \in C^s S^{2m}_{1,0}$ be an $L \times L$ system. Assume $s > m$ and*

$$p(x,\xi) + p(x,\xi)^* \ge C|\xi|^{2m} I, \quad \text{for } |\xi| \ge K. \tag{2.4.7}$$

Then, for any $\mu < m$, there exist C_1 and C_2 such that

$$Re\ (p(x,D)u,u) \ge C_1 \|u\|^2_{H^m} - C_2 \|u\|^2_{H^\mu}. \tag{2.4.8}$$

PROOF: It suffices to establish (2.4.8) for some $\mu < m$. Use the decomposition (2.4.1), with $p^\# \in S^{2m}_{1,\delta}$, $p^b \in C^{s-t} S^{2m-\delta t}_{1,0}$. For δ slightly less than 1, we can apply Gårding's inequality in its familiar form to $p^\#(x,D)$, and since, by Proposition 2.1.D, the perturbation $p^b(x,D)$ maps H^m to $H^{-m+\gamma}$, provided $s-t > m-\gamma$ and $\delta t > \gamma$, we obtain an estimate of the form (2.4.8).

Chapter 3: Paradifferential operators

The key tool of paradifferential operator calculus is developed in this chapter, beginning with Meyer's ingenious formula for $F(u)$ as $M(x, D)u + R$ where F is smooth in its argument(s), u belongs to a Hölder or Sobolev space, $M(x, D)$ is a pseudodifferential operator of type $1,1$, and R is smooth. From there, one applies symbol smoothing to $M(x, \xi)$ and makes use of results established in Chapter 2. The tool that arises is quite powerful in nonlinear analysis. The first glimpse we give of this is that it automatically encompasses some important Moser estimates. We re-derive elliptic regularity results established in Chapter 2, after establishing some microlocal regularity results. In §3.3 we do this using symbol smoothing with $\delta < 1$; in §3.4 we present some results of Bony and Meyer dealing with the $\delta = 1$ case, the case of genuine paradifferential operators.

In §3.5 some product estimates are established which, together with the operator calculus of §3.4, yield in §3.6 some useful commutator estimates, including important commutator estimates of Coifman and Meyer [CM] and of Kato and Ponce [KP]. We also discuss connections with the $T(1)$ Theorem.

§3.1. Composition and paraproducts

Following [M1], we discuss the connection between $F(u)$, for smooth nonlinear F, and the action on u of certain pseudodifferential operators of type 1,1. Let $\{\psi_j\}$ be the S_1^0 partition of unity (1.3.1), and set $\Psi_k(\xi) = \sum_{j \le k} \psi_j(\xi)$. Given u, e.g., in $C^r(\mathbb{R}^n)$, set

$$u_k = \Psi_k(D)u, \tag{3.1.1}$$

and write

$$F(u) = F(u_0) + [F(u_1) - F(u_0)] + \cdots + [F(u_{k+1}) - F(u_k)] + \cdots \tag{3.1.2}$$

Then write

$$\begin{aligned} F(u_{k+1}) - F(u_k) &= F(u_k + \psi_{k+1}(D)u) - F(u_k) \\ &= m_k(x)\psi_{k+1}(D)u, \end{aligned} \tag{3.1.3}$$

where

$$m_k(x) = \int_0^1 F'(\Psi_k(D)u + t\psi_{k+1}(D)u)dt \tag{3.1.4}$$

Consequently, we have

$$\begin{aligned} F(u) &= F(u_0) + \sum_{k=0}^{\infty} m_k(x)\psi_{k+1}(D)u \\ &= M(x, D)u + F(u_0) \end{aligned} \tag{3.1.5}$$

where

$$M(x,\xi)=\sum_{k=0}^{\infty} m_k(x)\psi_{k+1}(\xi)=M_F(u;x,\xi). \tag{3.1.6}$$

We claim

$$M(x,\xi)\in S^0_{1,1}, \tag{3.1.7}$$

provided u is continuous. To estimate $M(x,\xi)$, note first that, by (3.1.4),

$$\|m_k\|_{L^\infty}\le \sup|F'(\lambda)|. \tag{3.1.8}$$

To estimate higher derivatives, we use the classical estimate

$$\|D^\ell g(h)\|_{L^\infty}\le C\sum_{1\le\nu\le\ell}\|g'\|_{C^{\nu-1}}\|h\|_{L^\infty}^{\nu-1}\|D^\ell h\|_{L^\infty} \tag{3.1.9}$$

to obtain

$$\|D_x^\ell m_k\|_{L^\infty}\le C_\ell\|F''\|_{C^{\ell-1}}\langle\|u\|_{L^\infty}\rangle^{\ell-1}\cdot 2^{k\ell}, \tag{3.1.10}$$

granted the following estimates, which hold for all $u\in L^\infty$:

$$\|\Psi_k(D)u+t\psi_{k+1}(D)u\|_{L^\infty}\le C\|u\|_{L^\infty}, \tag{3.1.11}$$

and

$$\|D^\ell[\Psi_k(D)u+t\psi_{k+1}(D)u]\|_{L^\infty}\le C_\ell 2^{k\ell}\|u\|_{L^\infty} \tag{3.1.12}$$

for $t\in[0,1]$. Consequently, (3.1.6) yields

$$|D_\xi^\alpha M(x,\xi)|\le C_\alpha\sup_\lambda|F'(\lambda)|\langle\xi\rangle^{-|\alpha|} \tag{3.1.13}$$

and, for $|\beta|\ge 1$,

$$|D_x^\beta D_\xi^\alpha M(x,\xi)|\le C_{\alpha\beta}\|F''\|_{C^{|\beta|-1}}\langle\|u\|_{L^\infty}\rangle^{|\beta|-1}\langle\xi\rangle^{|\beta|-|\alpha|}. \tag{3.1.14}$$

We give a formal statement of the result just established.

PROPOSITION 3.1.A. *If F is C^∞ and $u\in C^r$ with $r\ge 0$, then*

$$F(u)=M_F(u;x,D)u+R(u) \tag{3.1.15}$$

where

$$R(u) = F(\psi_0(D)u) \in C^\infty$$

and

$$M_F(u; x, \xi) = M(x, \xi) \in S^0_{1,1}. \tag{3.1.16}$$

Applying Theorem 2.1.A, we have

$$\|M(x, D)f\|_{H^{s,p}} \le K\|f\|_{H^{s,p}} \tag{3.1.17}$$

for $p \in (1, \infty)$, $s > 0$, with

$$K = K_N(F, u) = C\|F'\|_{C^N}[1 + \|u\|^N_{L^\infty}], \tag{3.1.18}$$

provided $0 < s < N$, and similarly

$$\|M(x, D)f\|_{C^s_*} \le K\|f\|_{C^s_*}. \tag{3.1.19}$$

Using $f = u$, we have the following well known and important Moser-type estimates:

$$\|F(u)\|_{H^{s,p}} \le K_N(F, u)\|u\|_{H^{s,p}} + \|R(u)\|_{H^{s,p}}, \tag{3.1.20}$$

and

$$\|F(u)\|_{C^s_*} \le K_N(F, u)\|u\|_{C^s_*} + \|R(u)\|_{C^s_*}, \tag{3.1.21}$$

given $1 < p < \infty$, $0 < s < N$, with $K_N(F, u)$ as in (3.1.18); this involves the L^∞-norm of u, and one can use $\|F'\|_{C^N(I)}$ where I contains the range of u. Note that, if $F(u) = u^2$, then $F'(u) = 2u$, and higher powers of $\|u\|_{L^\infty}$ do not arise; hence we recover the familiar estimate

$$\|u^2\|_{H^{s,p}} \le C_s\|u\|_{L^\infty} \cdot \|u\|_{H^{s,p}}, \quad s > 0, \tag{3.1.22}$$

with a similar estimate on $\|u^2\|_{C^s_*}$.

It will be useful to have further estimates on the symbol $M(x, \xi) = M_F(u; x, \xi)$, when $u \in C^r$ with $r > 0$. The estimate (3.1.12) extends to

$$\begin{aligned} \big\|D^\ell\big[\Psi_k(D)f + t\psi_{k+1}(D)f\big]\big\|_{L^\infty} &\le C_\ell\|f\|_{C^r}, \quad \ell \le r, \\ &\ C_\ell 2^{k(\ell-r)}\|f\|_{C^r}, \quad \ell > r, \end{aligned} \tag{3.1.23}$$

so we have, when $u \in C^r$,

$$\begin{aligned} |D^\beta_x D^\alpha_\xi M(x, \xi)| &\le K_{\alpha\beta}\langle\xi\rangle^{-|\alpha|}, \quad |\beta| \le r, \\ &\ K_{\alpha\beta}\langle\xi\rangle^{-|\alpha|+|\beta|-r}, \quad |\beta| > r, \end{aligned} \tag{3.1.24}$$

with

$$K_{\alpha\beta} = K_{\alpha\beta}(F, u) = C_{\alpha\beta}\|F'\|_{C^{|\beta|}}[1 + \|u\|_{C^r}^{|\beta|}]. \tag{3.1.25}$$

Also, since $\Psi_k(D)+t\psi_{k+1}(D)$ is uniformly bounded on C^r, for $t \in [0,1]$, $k \geq 0$, we have

$$\|D_\xi^\alpha M(\cdot,\xi)\|_{C^r} \leq K_{\alpha r}\langle\xi\rangle^{-|\alpha|}, \tag{3.1.26}$$

where $K_{\alpha r}$ is as in (3.1.25), with $|\beta| = [r] + 1$. This last estimate shows that

$$u \in C^r \Longrightarrow M_F(u; x, \xi) \in C^r S^0_{1,0}. \tag{3.1.27}$$

This is useful additional information; for example (3.1.17) and (3.1.19) hold for $s > -r$, and of course we can apply the symbol smoothing of §1.3.

It will be useful to have terminology expressing the structure of the symbols we produce. Given $r \geq 0$, we say

(3.1.28)
$$p(x,\xi) \in \mathcal{A}^r S^m_{1,\delta} \Longleftrightarrow \|D_\xi^\alpha p(\cdot,\xi)\|_{C^r} \leq C_\alpha\langle\xi\rangle^{m-|\alpha|}$$
$$\text{and } |D_x^\beta D_\xi^\alpha p(x,\xi)| \leq C_{\alpha\beta}\langle\xi\rangle^{m-|\alpha|+\delta(|\beta|-r)}, \ |\beta| > r.$$

Thus (3.1.24) - (3.1.26) yield

$$M(x,\xi) \in \mathcal{A}^r S^0_{1,1} \tag{3.1.29}$$

for the $M(x,\xi)$ of Proposition 3.1.A. If $r \in \mathbb{R}^+ \setminus \mathbb{Z}^+$, the class $\mathcal{A}^r S^m_{1,1}$ coincides with the symbol class denoted by $\mathcal{A}^m_r$ by Meyer [M1]. Clearly $\mathcal{A}^0 S^m_{1,\delta} = S^m_{1,\delta}$, and

$$\mathcal{A}^r S^m_{1,\delta} \subset C^r S^m_{1,0} \cap S^m_{1,\delta}.$$

Also from the definition we see that

$$p(x,\xi) \in \mathcal{A}^r S^m_{1,\delta} \Longrightarrow D_x^\beta p(x,\xi) \in \begin{cases} S^m_{1,\delta} & \text{for } |\beta| \leq r \\ S^{m+\delta(|\beta|-r)}_{1,\delta} & \text{for } |\beta| \geq r. \end{cases} \tag{3.1.30}$$

It is also natural to consider a slightly smaller symbol class:

(3.1.31)
$$p(x,\xi) \in \mathcal{A}^r_0 S^m_{1,\delta} \Longleftrightarrow \|D_\xi^\alpha p(\cdot,\xi)\|_{C^{r+s}} \leq C_{\alpha s}\langle\xi\rangle^{m-|\alpha|+\delta s}, \quad s \geq 0.$$

Considering the cases $s = 0$ and $s = |\beta| - r$, we see that

$$\mathcal{A}_0^r S_{1,\delta}^m \subset \mathcal{A}^r S_{1,\delta}^m.$$

We also say

$$p(x,\xi) \in {}^r S_{1,\delta}^m \Longleftrightarrow \text{the right side of (3.1.30) holds ,} \tag{3.1.32}$$

so

$$\mathcal{A}^r S_{1,\delta}^m \subset {}^r S_{1,\delta}^m.$$

The following result refines (3.1.29).

PROPOSITION 3.1.B. *For the symbol $M(x,\xi) = M_F(u;x,\xi)$ of Proposition 3.1.A, we have*

$$M(x,\xi) \in \mathcal{A}_0^r S_{1,1}^m \tag{3.1.33}$$

provided $u \in C^r$, $r \geq 0$.

PROOF: For this, we need

$$\|m_k\|_{C^{r+s}} \leq C \cdot 2^{ks}. \tag{3.1.34}$$

Now, extending (3.1.9), we have

$$\|g(h)\|_{C^{r+s}} \leq C\|g\|_{C^N}[1 + \|h\|_{L^\infty}^N](\|h\|_{C^{r+s}} + 1) \tag{3.1.35}$$

with $N = [r + s] + 1$, as a consequence of (3.1.21) when $r + s$ is not an integer, and by (3.1.9) when it is. This gives, via (3.1.4),

$$\|m_k\|_{C^{r+s}} \leq C(\|u\|_{L^\infty}) \sup_{t \in I} \|(\Psi_k + t\psi_{k+1})u\|_{C^{r+s}} \tag{3.1.36}$$

where $I = [0,1]$. However,

$$\|(\Psi_k + t\psi_{k+1})u\|_{C^{r+s}} \leq C \cdot 2^{ks}\|u\|_{C^r}. \tag{3.1.37}$$

For $r + s \in \mathbb{Z}^+$, this follows from (1.3.11); for $r + s \notin \mathbb{Z}^+$ it follows as in the proof of Lemma 1.3.A, since

$$2^{-ks}\Lambda^s(\Psi_k + t\psi_{k+1}) \text{ is bounded in } OPS_{1,0}^0. \tag{3.1.38}$$

This establishes (3.1.34), and hence (3.1.33) is proved.

Returning to symbol smoothing, if we use the method of §1.3 to write

$$M(x,\xi) = M^{\#}(x,\xi) + M^{b}(x,\xi), \tag{3.1.39}$$

then (3.1.27) implies

$$M^{\#}(x,\xi) \in S^{m}_{1,\delta}, \quad M^{b}(x,\xi) \in C^{r}S^{m-r\delta}_{1,\delta}. \tag{3.1.40}$$

We now refine these results; for $M^{\#}$ we have a general result.

PROPOSITION 3.1.C. *For the symbol decomposition of §1.3,*

$$p(x,\xi) \in C^{r}S^{m}_{1,0} \Longrightarrow p^{\#}(x,\xi) \in \mathcal{A}^{r}_{0}S^{m}_{1,\delta}. \tag{3.1.41}$$

PROOF: This is a simple modification of Proposition 1.3.D, which essentially says $p^{\#}(x,\xi) \in \mathcal{A}^{r}S^{m}_{1,\delta}$; we simply supplement (1.3.11) with

$$\|J_{\epsilon}f\|_{C^{r+s}_{*}} \leq C\,\epsilon^{-s}\|f\|_{C^{r}_{*}}, \; s \geq 0, \tag{3.1.42}$$

which is basically the same as (3.1.37).

To treat $M^{b}(x,\xi)$, we have, for $\delta \leq \gamma$,

$$p(x,\xi) \in \mathcal{A}^{r}_{0}S^{m}_{1,\gamma} \Longrightarrow p^{b}(x,\xi) \in C^{r}S^{m-\delta r}_{1,\delta} \cap \mathcal{A}^{r}_{0}S^{m}_{1,\gamma} \subset S^{m-\delta r}_{1,\gamma}, \tag{3.1.43}$$

where containment in $C^{r}S^{m-\delta r}_{1,\delta}$ follows from Proposition 1.3.E. To see the last inclusion, note that for $p^{b}(x,\xi)$ to belong to the intersection above implies

$$\|D^{\alpha}_{\xi}p^{b}(\cdot,\xi)\|_{C^{s}} \leq \begin{cases} C\langle\xi\rangle^{m-|\alpha|-\delta r+\delta s} & \text{for } 0 \leq s \leq r \\ C\langle\xi\rangle^{m-|\alpha|+(s-r)\gamma} & \text{for } s \geq r. \end{cases} \tag{3.1.44}$$

In particular these estimates imply $p^{b}(x,\xi) \in S^{m-r\delta}_{1,\gamma}$. This proves:

PROPOSITION 3.1.D. *For the symbol $M(x,\xi) = M_{F}(u;x,\xi)$ with decomposition* (3.1.39),

$$u \in C^{r} \Longrightarrow M^{b}(x,\xi) \in S^{-r\delta}_{1,1}. \tag{3.1.45}$$

We now discuss a few consequences of making the decomposition (3.1.39). Note that

$$u \in C^{r} \cap H^{s,p} \Longrightarrow F(u) = M^{\#}(x,D)u + R, \text{ with } R \in H^{s+r\delta,p}, \tag{3.1.50}$$

provided $r, s > 0$. If we pick $\delta \in (0,1)$, using the good algebraic, hence microlocal, properties of $OPS^m_{1,\delta}$, we have the following extension of Rauch's Lemma.

PROPOSITION 3.1.E. *If* $u \in C^r \cap H^{s,p}$, $r, s > 0$, $p \in (1, \infty)$, *then*

$$u \in H^{\sigma,p}_{mcl}(\Gamma) \Longrightarrow F(u) \in H^{\sigma,p}_{mcl}(\Gamma) \tag{3.1.51}$$

provided

$$s \leq \sigma < s + r. \tag{3.1.52}$$

In (3.1.51), Γ is a closed conic subset of $T^*\mathbb{R}^n \setminus 0$, and the meaning of the hypothesis

$$u \in H^{\sigma,p}_{mcl}(\Gamma)$$

is that there exists

$$A(x, D) \in OPS^0, \text{ elliptic on } \Gamma, \text{ such that } A(x, D)u \in H^{\sigma,p}.$$

For the proof, note that

$$B(x, D)F(u) = B(x, D)M^\#(x, D)u + B(x, D)R;$$

if $B(x, D) \in OPS^0$, then $B(x, D)R \in H^{s+r\delta,p}$, by (3.1.50).

Using $F(u) = u^2$, it follows that $C^r \cap H^{s,p} \cap H^{\sigma,p}_{mcl}(\Gamma)$ is an algebra, granted (3.1.52). It has been typical to establish this result in case $s = \frac{n}{p} + r$, which implies $H^{s,p} \subset C^r$ (if $r \notin \mathbb{Z}$), but there may be an advantage to the more general formulation given above.

Results discussed above extend easily to the case of a function F of several variables, say $u = (u_1, \dots, u_L)$. Directly extending (3.1.2) - (3.1.6), we have

$$F(u) = \sum_{j=1}^{L} M_j(x, D)u_j + F(\Psi_0(D)u) \tag{3.1.53}$$

with

$$M_j(x, \xi) = \sum_k m^j_k(x)\psi_{k+1}(\xi) \tag{3.1.54}$$

where

$$m^j_k(x) = \int_0^1 (\partial_j F)(\Psi_k(D)u + t\psi_{k+1}(D)u)dt. \tag{3.1.55}$$

Clearly the results established above apply to the $M_j(x,\xi)$ here, e.g.,

$$u \in C^r \Longrightarrow M_j(x,\xi) \in \mathcal{A}_0^r S_{1,1}^m. \tag{3.1.56}$$

In the particular case $F(u,v) = uv$, we obtain

$$uv = A(u;x,D)v + A(v;x,D)u + \Psi_0(D)u \cdot \Psi_0(D)v \tag{3.1.57}$$

where

$$A(u;x,\xi) = \sum_{k=1}^{\infty} \Big[\Psi_k(D)u + \frac{1}{2}\psi_{k+1}(D)u\Big]\psi_{k+1}(\xi). \tag{3.1.58}$$

Since this symbol belongs to $S_{1,1}^0$ for $u \in L^\infty$, we obtain the following well known extension of (3.1.22):

$$\|uv\|_{H^{s,p}} \le C\big[\|u\|_{L^\infty}\|v\|_{H^{s,p}} + \|u\|_{H^{s,p}}\|v\|_{L^\infty}\big], \tag{3.1.59}$$

for $s > 0,\ 1 < p < \infty$.

§3.2. Various forms of paraproduct

A linear operator related to the operator $M(x,D)$ of Proposition 3.1.A is the paraproduct, used in [Bo], [M1]. There are several versions of the paraproduct; one is

$$\pi(a,f) = \sum_{k\ge 1} (\Psi_{k-1}(D)a)(\psi_{k+1}(D)f). \tag{3.2.0}$$

Note that this is a special case of the symbol smoothing of §1.3, in which $\delta = 1$. In particular, we have the following.

PROPOSITION 3.2.A. *If $a \in C^r$, then*

$$af = \pi(a,f) + \rho_a(x,D)f \tag{3.2.1}$$

with

$$\rho_a(x,\xi) \in C^r S_{1,1}^{-r}. \tag{3.2.2}$$

Hence, for $p \in (1,\infty)$,

$$\rho_a(x,D) : H^{s-r,p} \longrightarrow H^{s,p},\ 0 < s < r. \tag{3.2.3}$$

Note that this result does not imply that $\pi(f,f)$ is a particularly good approximation to f^2; this point will be clarified below; see (3.2.13).

We will also use the notation

$$\pi(a,f) = \pi_a(x,D)f = T_a f, \tag{3.2.4}$$

the latter notation being due to Bony [Bo]. We want to compare $M(x,\xi) = \sum_{k=0}^{\infty} m_k(x)\psi_{k+1}(\xi)$, given by (3.1.4) - (3.1.6), with

$$\pi_{F'}(x,\xi) = \sum_{k=0}^{\infty} \tilde{m}_k(x)\psi_{k+1}(\xi),$$

where

$$\tilde{m}_k(x) = \Psi_{k-1}(D)F'(f). \tag{3.2.5}$$

Comparing this with

$$m_k(x) = \int_0^1 F'(\Psi_k(D)f + t\psi_{k+1}(D)f)dt \tag{3.2.6}$$

gives the following.

PROPOSITION 3.2.B. *If $f \in C^r$, $M(x,\xi)$ given by (3.1.4) - (3.1.6), then*

$$M(x,\xi) - \pi_{F'}(x,\xi) \in S^{-r}_{1,1}. \tag{3.2.7}$$

PROOF: What is needed is the estimate

$$|D_x^\beta(m_k - \tilde{m}_k)| \le C_\beta \cdot 2^{-rk+|\beta|k}, \tag{3.2.8}$$

given $f \in C^r$, which follows from (3.2.5) - (3.2.6).

In order to establish estimates of the form (3.2.8), we use the fact that, for $r \in \mathbb{R}^+ \setminus \mathbb{Z}^+$,

$$\begin{aligned} g \in C^r &\Longleftrightarrow \|\psi_k(D)g\|_{L^\infty} \le C \cdot 2^{-kr} \\ &\Longrightarrow \|(1 - \Psi_k(D))g\|_{L^\infty} \le C \cdot 2^{-kr}. \end{aligned} \tag{3.2.9}$$

Thus, for F smooth,

$$\begin{aligned} f \in C^r &\Longrightarrow \|F'(f) - \Psi_k(D)F'(f)\|_{L^\infty} \le C \cdot 2^{-kr} \\ &\text{and } \|F'(f) - F'(\Psi_k(D)f)\|_{L^\infty} \le C \cdot 2^{-kr}, \end{aligned} \tag{3.2.10}$$

giving the case $\beta = 0$ of (3.2.8). If we write

$$\begin{aligned} \Psi_k(D)F(f) - F(\Psi_k(D)f) = \Psi_k(D)\big(F(f) - F(\Psi_k(D)f)\big) \\ - (1 - \Psi_k(D))F(\Psi_k(D)f), \end{aligned} \tag{3.2.11}$$

and use

$$\|D_x^\beta \Psi_k(D)g\|_{C^r} \le C_\beta 2^{k|\beta|}\|g\|_{C^r}, \tag{3.2.12}$$

the rest of (3.2.8) easily follows.
Remark. (3.2.8) is related to Hörmander's Prop. 8.6.13 in [H4]. Furthermore, one can take $f \in C_*^r$.

In view of the identity (3.1.5), we have:

PROPOSITION 3.2.C. *If* $f \in C^r \cap H^{s,p}$, $r, s > 0$, $p \in (1, \infty)$, *then*

$$F(f) = \pi(F'(f), f) + R, \quad R \in H^{s+r,p}. \tag{3.2.13}$$

Taking $F(f) = f^2$, we see that a good approximation to f^2 is $2\pi(f, f)$.

We can also treat functions of several variables, as in (3.1.53) - (3.1.54). Thus, if $f = (f_1, \dots, F_L)$,

$$F(f) = \sum \pi((\partial_j F)(f), f_j) + R, \tag{3.2.14}$$

where $f \in C^r \cap H^{s,p} \Longrightarrow R \in H^{s+r,p}$.

There are a number of variants of the paraproduct (3.2.0). For example, picking $P, Q \in \mathcal{S}(\mathbb{R}^n)$ with $P(0) = 1,\ Q(0) = 0$, one can consider

$$\pi(a, f) = \int_0^\infty Q(tD)(P(tD)a \cdot Q(tD)f)\, t^{-1} dt, \tag{3.2.15}$$

which differs little from (3.2.0). With a bit more effort, one can use the simpler-looking expression

$$\pi(a, f) = \int_0^\infty P(tD)a \cdot Q(tD)f\ t^{-1} dt. \tag{3.2.16}$$

A generalization of (3.2.16) is used in (5.9) of Hörmander [H3] to produce a variant of the Nash-Moser implicit function theorem.

Another variant of the paraproduct is

$$\pi(a, f) = a_\chi(x, D)f \tag{3.2.17}$$

with

$$\hat{a}_\chi(\eta,\xi) = \hat{a}(\eta)\chi(\eta,\xi), \tag{3.2.18}$$

where we choose $\chi \in C^\infty(\mathbb{R}^n \times \mathbb{R}^n)$, homogeneous of degree 0 outside a compact set, such that

$$\begin{aligned} \chi(\eta,\xi) &= 0 \text{ for } |\eta| > \frac{1}{2}|\xi| \\ &\quad 1 \text{ for } |\eta| < \frac{|\xi|}{16} \text{ and } |\xi| > 2. \end{aligned} \tag{3.2.19}$$

One has $a_\chi(x,\xi) \in S^0_{1,1}$, and more precisely

$$\begin{aligned} |D_\xi^\alpha D_x^\beta a_\chi(x,\xi)| \le C_{\alpha\beta}\langle\xi\rangle^{-|\alpha|}, &\quad |\beta| < r \\ C_{\alpha\beta}\langle\xi\rangle^{|\beta|-|\alpha|-r}, &\quad |\beta| \ge r \end{aligned} \tag{3.2.20}$$

if $a \in C^r$. This formulation of paraproducts is the one mainly used by Bony [Bo]. It is proved in Lemma 4.1 of Hörmander [H2] that, in this case,

$$a_\chi(x,D) \in OPS^0_{1,1} \cap (OPS^0_{1,1})^*. \tag{3.2.21}$$

Also Hörmander has results on compositions of operators, in Theorem 6.4 of [H2]. See also Prop. 10.2.2 of [H4].

From (2.1.15) easily follows the estimate

$$\|\pi(a,f)\|_{H^{s,p}} \le C_{sp}\|a\|_{L^\infty}\|f\|_{H^{s,p}}, \tag{3.2.22}$$

for $p \in (1,\infty)$, $s \in \mathbb{R}$, at least when the form (3.2.0) is used. It follows from Theorem 33 of [CM] that, for $p \in (1,\infty)$,

$$\|\pi(a,f)\|_{L^p} \le C_p\|a\|_{L^p}\|f\|_{BMO}, \tag{3.2.23}$$

an estimate which suggests emphasizing the role of f as a multiplier rather than a. In fact, a notational switch, with f and a interchanged in the definition of $\pi(a,f)$, is frequently seen. This other convention was the one originally used in Coifman-Meyer [CM]. The estimate (3.2.23) will play an important role in §3.5; a proof is sketched in Appendix D.

§3.3. Nonlinear PDE and paradifferential operators

If F is smooth in its arguments, in analogy with (3.1.53) - (3.1.55) we have

$$F(x,D^m u) = \sum_{|\alpha|\le m} M_\alpha(x,D)D^\alpha u + F(x,D^m\Psi_0(D)u), \tag{3.3.1}$$

where $F(x, D^m \Psi_0(D)u) \in C^\infty$ and

$$M_\alpha(x,\xi) = \sum_k m_k^\alpha(x)\psi_{k+1}(\xi) \tag{3.3.2}$$

with

$$m_k^\alpha(x) = \int_0^1 (\partial F/\partial \zeta_\alpha)(\Psi_k(D)D^m u + t\psi_{k+1}(D)D^m u)dt. \tag{3.3.3}$$

As in Proposition 3.1.A and Proposition 3.1.B we have, for $r \geq 0$,

$$u \in C^{m+r} \Longrightarrow M_\alpha(x,\xi) \in \mathcal{A}_0^r S_{1,1}^0 \subset S_{1,1}^0 \cap C^r S_{1,0}^0. \tag{3.3.4}$$

In other words, if we set

$$M(u;x,D) = \sum_{|\alpha|\leq m} M_\alpha(x,D)D^\alpha, \tag{3.3.5}$$

we obtain

PROPOSITION 3.3.A. *If $u \in C^{m+r}$, $r \geq 0$, then*

$$F(x, D^m u) = M(u;x,D)u + R \tag{3.3.6}$$

with $R \in C^\infty$ and

$$M(u;x,\xi) \in \mathcal{A}_0^r S_{1,1}^m \subset S_{1,1}^m \cap C^r S_{1,0}^m. \tag{3.3.7}$$

Decomposing each $M_\alpha(x,\xi)$, we have as in (3.1.39) - (3.1.45),

$$M(u;x,\xi) = M^\#(x,\xi) + M^b(x,\xi) \tag{3.3.8}$$

with

$$M^\#(x,\xi) \in \mathcal{A}_0^r S_{1,\delta}^m \subset S_{1,\delta}^m \tag{3.3.9}$$

and

$$M^b(x,\xi) \in C^r S_{1,\delta}^{m-\delta r} \cap \mathcal{A}_0^r S_{1,1}^m \subset S_{1,1}^{m-r\delta}. \tag{3.3.10}$$

Let us explicitly recall that (3.3.9) implies

$$D_x^\beta M^\#(x,\xi) \in \begin{cases} S_{1,\delta}^m, & |\beta| \leq r, \\ S_{1,\delta}^{m+\delta(|\beta|-r)}, & |\beta| \geq r. \end{cases} \tag{3.3.11}$$

Note that the linearization of $F(x, D^m u)$ at u is given by

$$Lv = \sum_{|\alpha|\leq m} \tilde{M}_\alpha(x) D^\alpha v, \tag{3.3.12}$$

where

$$\tilde{M}_\alpha(x) = (\partial F/\partial \zeta_\alpha)(x, D^m u). \tag{3.3.13}$$

Comparison with (3.3.1) - (3.3.3) gives (for $u \in C^{m+r}$)

$$M(u; x, \xi) - L(x, \xi) \in C^r S^{m-r}_{1,1}, \tag{3.3.14}$$

by the same analysis as in the proof of the $\delta = 1$ case of (1.3.19). More generally, the difference in (3.3.14) belongs to $C^r S^{m-r\delta}_{1,\delta}$, $0 \leq \delta \leq 1$. Thus $L(x, \xi)$ and $M(u; x, \xi)$ have many qualitative properties in common.

In particular, given $u \in C^{m+r}$, the operator $M^\#(x, D) \in OPS^m_{1,\delta}$ is microlocally elliptic in any direction $(x_0, \xi_0) \in T^*\mathbb{R}^n \setminus 0$ which is noncharacteristic for $F(x, D^m u)$, which by definition means noncharacteristic for L. Now if

$$F(x, D^m u) = f, \tag{3.3.15}$$

and if $A \in OPS^0$ is microlocally supported near (x_0, ξ_0) and $Q \in OPS^{-m}_{1,\delta}$ is a microlocal parametrix for $M^\#(x, D)$ near (x_0, ξ_0), we have

$$Au = AQ(f - M^b(x, D)u), \quad \text{mod } C^\infty. \tag{3.3.16}$$

By (3.3.10) we have

$$AQM^b(x, D) : H^{m-r\delta+s,p} \longrightarrow H^{m+s,p}, \ s > 0. \tag{3.3.17}$$

(In fact $s > -(1-\delta)r$ suffices.) This gives the folowing microlocal regularity theorem.

THEOREM 3.3.B. *Suppose $u \in C^{m+r}$ satisfies (3.3.15), for which (x_0, ξ_0) is noncharacteristic. Then, for any $s > 0$, $p \in (1, \infty)$, $\delta \in (0, 1)$,*

$$u \in H^{m-r\delta+s,p}, \ f \in H^{s,p}_{mcl}(x_0, \xi_0) \Longrightarrow u \in H^{m+s,p}_{mcl}(x_0, \xi_0). \tag{3.3.18}$$

If $u \in C^{m+r}$ solves (3.3.15) in the elliptic case, where every direction is noncharacteristic, we can deduce from (3.3.18) that

$$u \in H^{m-\delta r+s,p}, \ f \in H^{s,p} \Longrightarrow u \in H^{m+s,p}, \tag{3.3.19}$$

granted $r > 0,\ s > 0,\ p \in (1, \infty)$. This sort of implication can be iterated, leading to the following re-proof of Theorem 2.2.G.

THEOREM 3.3.C. *Suppose, given $r > 0,\ u \in C^{m+r}$ satisfies (3.3.15) and this PDE is elliptic. Then, for each $s > 0,\ p \in (1, \infty)$,*

$$f \in H^{s,p} \Longrightarrow u \in H^{m+s,p} \text{ and } f \in C_*^s \Longrightarrow u \in C_*^{m+s}. \tag{3.3.20}$$

By way of further comparison with the methods of §2.2, we now re-derive Theorem 2.2.E, a regularity result for solutions to a quasi-linear elliptic PDE. Note that, in the quasi-linear case,

$$F(x, D^m u) = \sum_{|\alpha| \le m} a_\alpha(x, D^{m-1}u) D^\alpha u = f, \tag{3.3.21}$$

the construction above gives $F(x, D^m u) = M(u; x, D)u + R_0(u)$ with the following properties:

$$\begin{aligned} &u \in C^{m+r}\ (r \ge 0) \Longrightarrow \\ &\quad M(u; x, \xi) \in C^{r+1} S_{1,0}^m \cap S_{1,1}^m + C^r S_{1,0}^{m-1} \cap S_{1,1}^{m-1}. \end{aligned} \tag{3.3.22}$$

Of more interest to us now is that, for $0 < r < 1$,

$$u \in C^{m-1+r} \Longrightarrow M(u; x, \xi) \in C^r S_{1,0}^m \cap S_{1,1}^m + S_{1,1}^{m-r}, \tag{3.3.23}$$

which follows from (3.1.23). Thus we can decompose the term in $C^r S_{1,0}^m \cap S_{1,1}^m$ as in §1.3 and throw the term in $S_{1,1}^{m-r}$ into the remainder, to get

$$M(u; x, \xi) = M^\#(x, \xi) + M^b(x, \xi) \tag{3.3.24}$$

with

$$M^\#(x, \xi) \in S_{1,\delta}^m, \quad M^b(x, \xi) \in S_{1,1}^{m-r\delta}. \tag{3.3.25}$$

If $P(x, D) \in OPS_{1,\delta}^{-m}$ is a parametrix for the elliptic operator $M^\#(x, D)$, then whenever $u \in C^{m-1+r} \cap H^{m-1+\rho,p}$ is a solution to (3.3.21), we have, mod C^∞,

$$u = P(x, D)f - P(x, D)M^b(x, D)u. \tag{3.3.26}$$

Now

$$P(x, D)M^b(x, D) : H^{m-1+\rho,p} \longrightarrow H^{m-1+\rho+r\delta,p}, \text{ if } r + \rho > 1, \tag{3.3.27}$$

by the last part of (3.3.25). As long as this holds, we can iterate this argument, and obtain Theorem 2.2.E, with a shorter proof than given in §2.2.

More generally, consider

$$F(x, D^m u) = \sum_{|\alpha| \le m} a_\alpha(x, D^j u) D^\alpha u, \tag{3.3.28}$$

with $0 \le j < m$. We see that the conclusion of (3.3.23) holds if $u \in C^{j+r}$, and then the arguments yielding (3.3.24) - (3.3.27) continue to hold. Hence, in this case, one has regularity theorems assuming a priori that, with $p > 1$, $r > 0$,

$$u \in C^{j+r} \cap H^{m-1+\rho,p}, \ r + \rho > 1. \tag{3.3.29}$$

Under this hypothesis, we conclude that $f \in H^{s,q} \Longrightarrow u \in H^{s+m,q}$, etc.

In Bony's analysis of $F(x, D^m u)$, in [Bo], he used, in place of (3.3.1) - (3.3.3) the paraproduct approximation:

$$F(x, D^m u) = \sum_{|\alpha| \le m} \pi\big((\partial F/\partial \zeta_\alpha)(x, D^m u), D^\alpha u\big) + R, \tag{3.3.30}$$

where

$$u \in C^{m+r} \Longrightarrow R \in C^{2r}, \tag{3.3.31}$$

and $\pi(a, f)$ is as in one of the 3 definitions of §3.2.

In addition to paraproducts, Bony considered paradifferential operators, which can be defined as follows. Let $p(x, \xi) \in C^r S^m_{1,0}$. Then

$$T_p u(x) = p_\chi(x, D) u \tag{3.3.32}$$

where, in analogy with (3.2.18) - (3.2.19),

$$\hat{p}_\chi(\eta, \xi) = \chi(\eta, \xi) \hat{p}(\eta, \xi). \tag{3.3.33}$$

Then

$$p_\chi(x, D) \in OPS^m_{1,1} \cap (OPS^m_{1,1})^* = OP\tilde{S}^m_{1,1}, \tag{3.3.34}$$

as in (3.2.21). In fact, in Prop. 10.2.2 of [H4] it is shown that

$$\begin{aligned} D_x^\beta p_\chi(x, \xi) \in \ & \tilde{S}^m_{1,1} \quad |\beta| < r, \\ & \tilde{S}^{m-r+|\beta|}_{1,1}, \ |\beta| > r. \end{aligned} \tag{3.3.35}$$

We make a parenthetical comment. As noted above, we can write

$$F(x, D^m u) = M(u; x, D)u + R_0(u) \tag{3.3.36}$$

with $R_0(u) \in C^\infty$ and $M(u; x, D) \in OPS^m_{1,1}$ if $u \in C^m$. Consequently, the operator norm of $M(u; x, D)$ in $\mathcal{L}(H^{m+s,p}, H^{s,p})$, $1 < p < \infty$, depends on $\|u\|_{C^m}$, for $s > 0$, while for $s = 0$ it seems to depend on $\|u\|_{C^{m+r}}$ for some $r > 0$. An improvement on this is given in Proposition 3.5.G.

§3.4. Operator algebra

The operators $M(u; x, D) \in OPS^m_{1,1}$ which arose in §3.1 and §3.3 are not as well behaved as one would like under composition on the left by pseudodifferential operators. That is why decomposition into $M^\#(x, D) + M^b(x, D)$ was useful. The applications we have made so far have involved such a decomposition, defined in §1.3, choosing $\delta < 1$, so that $M^\#(x, D) \in OPS^m_{1,\delta}$ has a convenient symbol calculus. The remainder term $M^b(x, D)$ belongs to $OPS^{m-r\delta}_{1,1}$, and one despairs of doing anything with it except utilizing boundedness properties on various function spaces.

As noted, the paraproduct, defined by (3.2.0), is also an example of the construction of $M^\#(x, \xi)$ by symbol smoothing, this time with $\delta = 1$. Bony [Bo] and Meyer [M1] made use of the fact that $M^\#(x, D) \in OPS^m_{1,1}$ has a special property that allows a bit of symbol calculus to carry through. Though the algebraic structure on such $M^\#(x, D)$ is less well behaved than in the $\delta < 1$ case, one has the advantage that the recalcitrant remainder term $M^b(x, D)$ belongs to $OPS^{m-r}_{1,1}$, hence has a (slightly) lower order than one achieves by using symbol smoothing with $\delta < 1$.

The special property possessed by $M^\#(x, D)$ when it is a paradifferential operator, of the form (3.2.0) or more generally (3.3.30), is that its symbol belongs to the class $\mathcal{B}^r S^m_{1,1}$, defined as follows:

$$\begin{aligned} p(x,\xi) \in \mathcal{B}^r S^m_{1,1} \Longleftrightarrow p(x,\xi) &\in \mathcal{A}^r S^m_{1,1} \text{ and} \\ \hat{p}(\eta, \xi) &\text{ is supported in } |\eta| < |\xi|/10. \end{aligned} \tag{3.4.1}$$

Here $\hat{p}(\eta, \xi) = \int p(x, \xi) e^{-ix\cdot\eta} dx$. Thus the paradifferential operator construction writes an operator $M(u; x, D)$ of the form (3.3.6) as a sum

$$M(u; x, D) = M^\#(x, D) + M^b(x, D) \tag{3.4.2}$$

with

$$M^\#(x, \xi) \in \mathcal{B}^r S^m_{1,1}, \quad M^b(x, \xi) \in S^{m-r}_{1,1}. \tag{3.4.3}$$

If $r \in \mathbb{R}^+ \setminus \mathbb{Z}^+$, the class $\mathcal{B}^r S^m_{1,1}$ coincides with the symbol class denoted $\mathcal{B}^m_r$ by Meyer [M1].

We now present a couple of results of Meyer [M1], [M2], on products $p(x,D)q(x,D)$ when $q(x,\xi) \in \mathcal{B}^r S^m_{1,1}$.

THEOREM 3.4.A. *Assume* $r > 0$, $p(x,\xi) \in S^{\mu}_{1,1}$, *and* $q(x,\xi) \in \mathcal{B}^r S^m_{1,1}$. *Then*

$$p(x,D)q(x,D) = P(x,D) + R(x,D) \tag{3.4.4}$$

with

$$P(x,\xi) = \sum_{|\alpha|\le r} (i^{-|\alpha|}/\alpha!)\partial_\xi^\alpha p(x,\xi)\cdot \partial_x^\alpha q(x,\xi), \tag{3.4.5}$$

and

$$R(x,\xi) \in S^{m+\mu-r}_{1,1}. \tag{3.4.6}$$

We take (3.4.5) as the definition of $P(x,\xi)$, and are required to demonstrate (3.4.6). In particular, we need to show that

$$|R(x,\xi)| \le C\langle\xi\rangle^{m+\mu-r}. \tag{3.4.7}$$

We begin with the exact formula

(3.4.8)

$$R(x,\xi) = (2\pi)^{-n}\int \Big[p(x,\xi+\eta) - \sum_{|\alpha|\le r}(\eta^\alpha/\alpha!)\partial_\xi^\alpha p(x,\xi)\Big] e^{ix\cdot\eta}\hat{q}(\eta,\xi)\,d\eta.$$

Using the partition of unity $1 = \sum_{j=0}^{\infty}\psi_j(\eta)$ of (1.3.1), with ψ_0 supported on $|\eta| < 1$ and, for $j \ge 1$, ψ_j on $\frac{1}{3}2^j < |\eta| < 2^j$, insert this partition into the integrand on the right side of (3.4.8), to obtain

$$R(x,\xi) = \sum_{j=0}^{\infty} R_j(x,\xi). \tag{3.4.9}$$

To estimate the terms R_j, write

$$R_j(x,\xi) = \int \hat{A}_j(\eta)\hat{B}_j(\eta)\,d\eta, \tag{3.4.10}$$

where for convenience we have suppressed the variables x,ξ on the right, and set

$$\begin{aligned} \hat{A}_j(\eta) &= (2\pi)^{-n}\Big[p(x,\xi+\eta) - \sum_{|\alpha|\le r}(\eta^\alpha/\alpha!)\partial_\xi^\alpha p(x,\xi)\Big]\varphi_j(\eta), \\ \hat{B}_j(\eta) &= \hat{q}(\eta,\xi)\varphi_j(\eta)e^{ix\cdot\eta}, \end{aligned} \tag{3.4.11}$$

assuming $\psi_j(\eta) = \varphi_j(\eta)^2$. Thus $R_j(x,\xi) = \int A_j(-y)B_j(y)dy$, so

$$|R_j(x,\xi)| \le \|A_j\|_{L^1}\|B_j\|_{L^\infty}. \tag{3.4.12}$$

Since $q(x,\xi)$ is C^r in x, we have

$$\|B_j\|_{L^\infty} \le C2^{-jr}\langle\xi\rangle^m \tag{3.4.13}$$

and since $\hat{q}(\eta,\xi)$ vanishes for $|\eta| \ge |\xi|/10$, we have

$$\|B_j\|_{L^\infty} = 0 \text{ for } 2^j \ge \frac{3}{10}|\xi|. \tag{3.4.14}$$

We claim that, for each $\nu = 0,1,2,\dots,$

$$\|A_j\|_{L^1} \le C_\nu 2^{j(\nu+1)}\langle\xi\rangle^{\mu-\nu-1}. \tag{3.4.15}$$

Granted this, pick ν such that $\nu \le r < \nu+1$. Then (3.4.13) - (3.4.15) yield

$$\begin{aligned} |R(x,\xi)| &\le C\sum_{j=0}^{k} 2^{j(\nu+1-r)}\langle\xi\rangle^{m+\mu-\nu-1} \\ &\le C'2^{k(\nu+1-r)}\langle\xi\rangle^{m+\mu-\nu-1}, \end{aligned} \tag{3.4.16}$$

where k is the largest integer such that $2^k \le \frac{3}{10}|\xi|$. This yields the desired estimate (3.4.7). Estimates on $D_x^\beta D_\xi^\alpha R(x,\xi)$ are obtained in the same fashion, proving the theorem.

It remains to establish (3.4.15), for arbitrary $p(x,\xi) \in S^\mu_{1,1}$. This can be done using

$$\|A_j\|_{L^1} = \|\hat{A}_j\|_{\mathcal{F}L^1} \le C\|\hat{A}_j\|_{H^s(\mathbb{R}^n)}, \quad s > \frac{n}{2}, \tag{3.4.17}$$

the last inequality being one version of a Sobolev imbedding theorem. Estimating η-derivatives of the right side of the formula for $\hat{A}_j(\eta)$ in (3.4.11) is a routine consequence of Taylor's formula with remainder, and (3.4.15) results.

Following [M2], we next construct a microlocal parametrix. As usual, we say $q(x,\xi) \in S^m_{1,1}$ is elliptic on a closed conic set Γ if $|q(x,\xi)| \ge C|\xi|^m$ on Γ, for $|\xi|$ large. It is clear that in such a case there exists $p_0(x,\xi) \in S^{-m}_{1,1}$, equal to $q(x,\xi)^{-1}$ on Γ.

THEOREM 3.4.B. *Let $\Gamma_1 \subset\subset \Gamma$ be conic sets in $T^*\mathbb{R}^n \setminus 0$. If $r > 0$ and $q(x,\xi) \in \mathcal{B}^r S^m_{1,1}$ is elliptic on Γ, there exists $p(x,\xi) \in S^{-m}_{1,1}$ such that*

$$p(x,D)q(x,D) = P(x,D) + R(x,D) \tag{3.4.18}$$

with

$$P(x,\xi) \in S^0_{cl}, \text{ elliptic on } \Gamma_1, \tag{3.4.19}$$

and

$$R(x,\xi) \in S^{-r}_{1,1}. \tag{3.4.20}$$

PROOF: Take any $P(x,\xi)$ of the form (3.4.19), with conic support in Γ, and let $p_0(x,\xi) = q(x,\xi)^{-1}P(x,\xi)$, for $|\xi|$ large, $p_0(x,\xi) \in S^{-m}_{1,1}$. Then Theorem 3.4.A applies to $p_0(x,D)q(x,D)$. By a straightforward and standard induction one can construct $p_j(x,\xi) \in S^{-m-j}_{1,1}$, $j \le [r]$, such that (3.4.18) holds for $p = p_0 + \cdots + p_{[r]}$.

We remark that $p(x,\xi)$ so constructed actually belongs to $\mathcal{A}^r S^{-m}_{1,1}$. Thus we can write

$$p(x,\xi) = p^{\#}(x,\xi) + p^b(x,\xi),\ p^{\#} \in \mathcal{B}^r S^{-m}_{1,1},\ p^b \in S^{-m-r}_{1,1} \tag{3.4.21}$$

and hence

$$p^{\#}(x,D)q(x,D) = P(x,D) + R_1(x,D) \tag{3.4.22}$$

with $P(x,D)$ as in (3.4.18) - (3.4.19), and $R_1(x,\xi) \in S^{-r}_{1,1}$.

We next establish a regularity result.

LEMMA 3.4.C. *If $q(x,\xi) \in \mathcal{B}^r S^m_{1,1}$ is elliptic on a conic set Γ and if $u \in H^{s,p}$ while $q(x,D)u \in H^{\sigma,p}$, then $u \in H^{\sigma+m,p}_{mcl}(\Gamma)$, provided $p \in (1,\infty)$, $s > m$, and*

$$0 < \sigma + m \le s + r. \tag{3.4.23}$$

PROOF: Taking $p(x,\xi) \in \mathcal{A}^r S^{-m}_{1,1}$, $P(x,\xi) \in S^0_{cl}$, as in Theorem 3.4.B, we have

$$P(x,D)u = p(x,D)(q(x,D)u) - R(x,D)u \in H^{\sigma+m,p}, \tag{3.4.24}$$

provided (3.4.23) holds, which gives the proof

Next we obtain a result on the extent to which an operator with symbol in $\mathcal{B}^r S^m_{1,1}$ is microlocal.

PROPOSITION 3.4.D. *If $q(x,\xi) \in \mathcal{B}^r S^m_{1,1}$ and $u \in H^{s,p} \cap H^{\sigma,p}_{mcl}(\Gamma)$, then $q(x,D)u \in H^{\sigma-m,p}_{mcl}(\Gamma)$, provided $s > m$, $s > 0$, and*

$$s \le \sigma \le s + r. \tag{3.4.25}$$

PROOF: Adding $K\langle\xi\rangle^m$, we can assume $q(x,\xi)$ is elliptic on Γ. Set $v = q(x,D)u \in H^{s-m,p}$, granted $m < s$. By Theorem 3.4.B and the comment following its proof, given conic $\Gamma_1 \subset\subset \Gamma$, there exists $p^{\#}(x,\xi) \in \mathcal{B}^r S^{-m}_{1,1}$ such that (3.4.22) holds, with $P(x,\xi) \in S^0_{cl}$ elliptic on Γ_1. Hence

$$p^{\#}(x,D)v = P(x,D)u + R_1(x,D)u \in H^{\sigma,p},$$

granted $s + r \geq \sigma$. Since $p^{\#}(x,\xi)$ is elliptic on Γ_1, Lemma 3.4.C applies, to complete the proof.

We now obtain an improvement of Lemma 3.4.C to the following microlocal regularity result.

PROPOSITION 3.4.E. *The assertion of Lemma 3.4.C holds with the hypothesis on $q(x,D)u$ weakened to $q(x,D)u \in H^{\sigma,p}_{mcl}(\Gamma)$.*

PROOF: In (3.4.24), we see now that $p(x,D)(q(x,D)u) \in H^{\sigma+m,p}_{mcl}(\Gamma)$, so therefore $P(x,D)u \in H^{\sigma+m,p}$, under the hypotheses of the Lemma.

We can use these propositions to sharpen up some of the results of §3.1 and §3.3. For example, using

$$F(u) = M^{\#}(x,D)u + R$$

with $M^{\#}(x,\xi) \in \mathcal{B}^r S^0_{1,1}$, $R \in H^{s+r,p}$, given $u \in C^r \cap H^{s,p}$, $r, s > 0$, we see that in the Rauch Lemma, Proposition 3.1.E, the condition (3.1.52) can be sharpened to $s \leq \sigma \leq s + r$. Similarly, in the microlocal regularity result on solutions to a nonlinear PDE, $F(x, D^m u) = f$, given in Theorem 3.3.B, one can sharpen (3.3.18) to

$$(3.4.26)\quad u \in H^{m-r+s,p} \cap C^{m+r},\ f \in H^{s,p}_{mcl}(x_0,\xi_0) \Longrightarrow u \in H^{m+s,p}_{mcl}(x_0,\xi_0).$$

However, this does not yield a sharpening of the elliptic regularity result of Theorem 3.3.C.

Whether these sharpenings constitute major or minor improvements might be regarded as a matter of taste. Material in following sections should demonstrate essential advantages of results on $\mathcal{B}^r S^m_{1,1}$ described above. In the next section we establish a commutator estimate which seems to be inaccessible by the tools used in §3.1 and §3.3, involving operator calculus in $OPS^m_{1,\delta}$, with $\delta < 1$. In later chapters, particularly Chapter 5, we use such an estimate to obtain results which are not only sharper than those available by techniques involving use of the latter calculus, but have a natural feel to them which should provide good evidence of the usefulness of the paradifferential calculus.

Though we have concentrated on the study of $\mathcal{B}^r S^m_{1,1}$ for $r > 0$, it is of interest to note the following.

PROPOSITION 3.4.F. *If $p(x,\xi) \in \mathcal{B}^0 S^m_{1,1}$ and $1 < p < \infty$, then*

$$p(x,D) : H^{s+m,p} \Longrightarrow H^{s,p} \text{ for all } s \in \mathbb{R}. \tag{3.4.27}$$

PROOF: Looking at the proof of Theorem 2.1.A, reduced to the study of $q(x,\xi) = q_1 + q_2 + q_3$ in (2.1.14), we see that it suffices to consider $q_1(x,D)$. This is done in (2.1.15).

Hörmander [H2] shows that operators with symbol in $\mathcal{B}^r S^m_{1,1}$ belong to $\tilde{\Psi}^m_{1,1} = OPS^m_{1,1} \cap (OPS^m_{1,1})^*$. This contains the results (3.2.21), (3.3.35), and (3.4.27). Theorem 3.4.A is generalized in [H2] to the case $q(x,D) \in \tilde{\Psi}^m_{1,1}$. See also Bourdaud [BG] for algebraic properties of $\tilde{\Psi}^m_{1,1}$.

The space $OP\mathcal{B}^r S^*_{1,1}$ does not quite form an algebra. However, it is easy to see that, if $p(x,\xi) \in \mathcal{B}^r S^\mu_{1,1}$ and $q(x,\xi) \in \mathcal{B}^r S^m_{1,1}$ have the further property of $\hat{p}(\eta,\xi)$, $\hat{q}(\eta,\xi)$ being supported on $|\eta| < |\xi|/30$, the the operators $P(x,D)$ and $R(x,D)$ in (3.4.4) satisfy

$$P(x,\xi) \in \mathcal{B}^0 S^{m+\mu}_{1,1}, \quad R(x,\xi) \in \mathcal{B}^0 S^{m+\mu-r}_{1,1}. \tag{3.4.28}$$

To see this, note that the symbol $p\#q$ of $p(x,D)q(x,D)$ is given by

$$(p\#q)(x,\xi) = (2\pi)^{-n} \int p(x,\xi+\zeta) e^{ix\cdot\zeta} \hat{q}(\zeta,\xi)\, d\zeta,$$

so

$$(p\#q)\hat{\ }(\eta,\xi) = \int \hat{p}(\eta-\zeta,\xi+\zeta)\hat{q}(\zeta,\xi)\, d\zeta.$$

Also note that, if we start with $p(x,\xi) \in \mathcal{B}^r S^\mu_{1,1}$, applying a cutoff to achieve the further restriction on $\hat{p}(\eta,\xi)$ alters $p(x,\xi)$ by an element of $\mathcal{B}^r S^{\mu-r}_{1,1}$. Thus from Theorem 3.4.A we have:

COROLLARY 3.4.G. *If $r > 0$, $p(x,\xi) \in \mathcal{B}^r S^\mu_{1,1}$, $q(x,\xi) \in \mathcal{B}^r S^m_{1,1}$, then*

$$p(x,D)q(x,D) = P_1(x,D) + R_1(x,D) \tag{3.4.28}$$

with $P_1(x,\xi) \in \mathcal{B}^0 S^{m+\mu}_{1,1}$, given mod $S^{m+\mu-r}_{1,1}$ by (3.4.5), and

$$R_1(x,D) : H^{s,p} \longrightarrow H^{s-m-\mu+r,p} \tag{3.4.30}$$

for all $s \in \mathbb{R}$, $p \in (1,\infty)$. Consequently, in the scalar case,

$$[p(x,D), q(x,D)] : H^{s,p} \longrightarrow H^{s-m-\mu+\sigma,p}, \quad \sigma = \min(r,1). \tag{3.4.31}$$

Having discussed operators with symbols in $\mathcal{B}^r S^m_{1,1}$, we now note some applications to a smaller class of operators, with symbols in a space denoted Σ^m_r. These operators are the ones used by Bony in [Bo]. By definition, Σ^m_r is the image of $C^r S^m_{cl}$ under a smoothing process, with $\delta = 1$, of the form

$$a^{\#}(\cdot,\xi) = \sum_{k \geq 5} \Psi_{k-5}(D)a(\cdot,\xi)\psi_{k+1}(\xi). \tag{3.4.32}$$

In case $a = a(x)$ is independent of ξ, then

$$a^{\#}(x, D)u = \pi(a, u) = T_a u \tag{3.4.33}$$

is paramultiplication, as in (3.2.0). If we have

$$a(x,\xi) = \sum a_j(x)\beta_j(\xi), \tag{3.4.34}$$

then

$$a^{\#}(x,\xi)u = \sum T_{a_j}\beta_j(D)u. \tag{3.4.35}$$

As in [Bo], we also denote this operator by T_a. we will also denote by T'_a the operator obtained by replacing $\Psi_{k-5}(D)$ by $\Psi_{k-10}(D)$ in (3.4.32). It is clear that, for $r \geq 0$,

$$a(x,\xi) \in C^r S^m_{cl} \Longrightarrow T_a \in OP\mathcal{B}^r S^m_{1,1},\ T_a - T'_a \in OP\mathcal{B}^r S^{m-r}_{1,1}. \tag{3.4.36}$$

In particular,

$$T_a - T'_a : H^{s,p} \longrightarrow H^{s-m+r,p} \tag{3.4.37}$$

for all $s \in \mathbb{R}$, $p \in (1, \infty)$. From Corollary 3.4.G we have:

COROLLARY 3.4.H. *If $a_j(x,\xi) \in C^r S^{m_j}_{cl}$, then*

$$T'_{a_1}T'_{a_2} - T'_{a_1 a_2} \in OP\mathcal{B}^r S^{m_1+m_2-\sigma}_{1,1},\quad \sigma = \min(r, 1). \tag{3.4.38}$$

Consequently,

$$T_{a_1}T_{a_2} - T_{a_1 a_2} : H^{s,p} \longrightarrow H^{s-m_1-m_2+\sigma,p} \tag{3.4.39}$$

for $s \in \mathbb{R}$, $p \in (1, \infty)$.

§3.5. Product estimates

There are results more sophisticated than (3.1.59) on products that can be obtained from a careful analysis of the terms in

$$fg = T_f g + T_g f + R(f,g), \tag{3.5.1}$$

where, as in (3.4.33),

$$T_f g = \sum_k \big(\Psi_{k-5}(D)f\big) \cdot \big(\psi_{k+1}(D)g\big), \tag{3.5.2}$$

and

$$R(f,g) = \sum_k \big(\psi_k^a(D)f\big) \cdot \big(\psi_k(D)g\big) \tag{3.5.3}$$

where

$$\psi_k^a(\xi) = \sum_{\ell=k-5}^{k+5} \psi_\ell(\xi). \tag{3.5.4}$$

Note that $R(f,g) = R(g,f)$.

A number of results on $T_f g$ and $T_g f$ follow from the obvious fact that

$$f \in L^\infty \Longrightarrow T_f \in OP\mathcal{B}^0 S^0_{1,1}. \tag{3.5.5}$$

Hence

$$\|T_f g\|_{H^{s,p}} \le C_{sp} \|f\|_{L^\infty} \|g\|_{H^{s,p}} \tag{3.5.6}$$

for $s \in \mathbb{R},\ p \in (1,\infty)$. Almost equally obvious is

$$f \in C_*^{-\mu} \Longrightarrow T_f \in OP\mathcal{B}^0 S^\mu_{1,1} \text{ if } \mu > 0, \tag{3.5.7}$$

so

$$\|T_f g\|_{H^{s-\mu,p}} \le C \|f\|_{C_*^{-\mu}} \|g\|_{H^{s,p}}, \quad \mu > 0, \tag{3.5.8}$$

for $s \in \mathbb{R},\ p \in (1,\infty)$. Since

$$H^{r,p} \subset C_*^{-\mu} \text{ for } \frac{n}{p} - r = \mu, \tag{3.5.9}$$

we have in particular

$$\|T_f g\|_{H^{r+s-\frac{n}{p},p}} \le C\|f\|_{H^{r,p}}\|g\|_{H^{s,p}}, \quad r < \frac{n}{p}, \tag{3.5.10}$$

for $s \in \mathbb{R},\ p \in (1,\infty)$.

As for $R(f,g) = R_f g$, we have in partial analogy to (3.5.5) and (3.5.7) that

$$f \in C_*^r \Longrightarrow R_f \in OPS_{1,1}^{-r}, \quad r \in \mathbb{R}, \tag{3.5.11}$$

so, in analogy to (3.5.6) and (3.5.8),

$$\|R(f,g)\|_{H^{s+r,p}} \le C\|f\|_{C_*^r}\|g\|_{H^{s,p}}, \quad r \in \mathbb{R},\ s > -r, \tag{3.5.12}$$

for $p \in (1,\infty)$, while, in analogy to (3.5.10), we have

$$\|R(f,g)\|_{H^{r+s-\frac{n}{p},p}} \le C\|f\|_{H^{r,p}}\|g\|_{H^{s,p}}, \quad r+s > \frac{n}{p}, \tag{3.5.13}$$

if $p \in (1,\infty)$.

More subtle results on R_f, as well as useful results on

$$T_f^\rho g = T_g f, \tag{3.5.14}$$

can be deduced from the following important result, Theorem 33 of [CM]:

THEOREM 3.5.A. *Let $\varphi, \psi \in \mathcal{S}(\mathbb{R}^n)$, $\psi(0) = 0$. (Slightly more general φ, ψ are handled in [CM].) Let $m \in L^\infty(\mathbb{R}^+)$. Then the operator*

$$\tau(a,f) = \int_0^\infty \big(\varphi(tD)f\big)\cdot\big(\psi(tD)a\big)\frac{m(t)}{t}\,dt \tag{3.5.15}$$

satisfies

$$\|\tau(a,f)\|_{L^2} \le C\|a\|_{BMO}\|f\|_{L^2}. \tag{3.5.16}$$

There is a corresponding estimate for the discrete analogue

$$\sum_k \big(\varphi(2^{-k}D)f\big)\cdot\big(\psi(2^{-k}D)a\big)m_k. \tag{3.5.17}$$

The operators $R(a,f)$ and $T_a^\rho f$ are special cases of this. Consequently,

$$\|R(f,g)\|_{L^2} \le C\|f\|_{BMO}\|g\|_{L^2} \tag{3.5.18}$$

and

$$\|T_f g\|_{L^2} \le C\|g\|_{BMO}\|f\|_{L^2}. \tag{3.5.19}$$

In Appendix D we establish a variant of Theorem 3.5.A, which is strong enough to yield the estimates (3.5.18) and (3.5.19).

Since $R_f \in OPS^0_{1,1}$ is a singular integral operator for $f \in BMO \subset C^0_*$, Calderon-Zygmund theory, discussed in §0.11, implies that (3.5.18) can be extended to

$$\|R(f,g)\|_{L^p} \le C_p\|f\|_{BMO}\|g\|_{L^p}, \quad p \in (1,\infty). \tag{3.5.20}$$

We produce an extension of this which will be useful.

PROPOSITION 3.5.B. *Let X^r be a Banach space with the property*

$$P \in OPS^r_{1,0} \Longrightarrow P : X^r \longrightarrow BMO. \tag{3.5.21}$$

Then, for $p \in (1,\infty)$, $r \in \mathbb{R}$,

$$\|R(f,g)\|_{L^p} \le C\|f\|_{X^r}\|g\|_{H^{-r,p}}. \tag{3.5.22}$$

PROOF: Decompose f into $\sum_{\ell=1}^{20} f_\ell$, via operators in $OPS^0_{1,0}$, so the Fourier transforms of f_ℓ lie in $2^k \le |\xi| \le 2^{k+2}$ with $k = \ell \bmod 20$. Similarly decompose g. It suffices to estimate each $R(f_\ell, g_m)$. In such a case, we can find

$$F_\ell = Q_+ f_\ell \in BMO, \quad G_m = Q_- g_m \in L^p, \quad Q_\pm \in OPS^{\pm r}_{1,0} \tag{3.5.23}$$

such that, for each k,

$$\psi^a_k(D) f_\ell = 2^{-kr}\psi^a_k(D)F_\ell, \quad \psi_k(D)g_m = 2^{kr}\psi_k(D)G_m. \tag{3.5.24}$$

Hence $R(f_\ell, g_m) = R(F_\ell, G_m)$, so the estimate (3.5.22) follows from (3.5.20).

Note that some examples of X^r satisfying (3.5.21) are

$$X^r = H^{\frac{n}{p}+r,p}, \ r \in \mathbb{R}; \quad X^k = C^k, \ k \in \mathbb{Z}^+. \tag{3.5.25}$$

To give one example of how (3.5.22) helps to sharpen conclusions using (3.5.12), note that, if we apply (3.5.10) and (3.5.13) to the decomposition (3.5.1) of fg, we conclude that

$$f \in H^{r,p}, \ g \in H^{s,p}, \quad r < \frac{n}{p}, \ s < \frac{n}{p} \tag{3.5.26}$$

implies

$$fg \in H^{r+s-\frac{n}{p},p} \text{ provided also } r+s > \frac{n}{p}. \tag{3.5.27}$$

The only term for which this extra condition is required is $R(f,g)$; the weakness is in (3.5.13). By Proposition 3.5.B and (3.5.25), we can extend this implication to the case $r+s=\frac{n}{p}$:

$$fg \in L^p \text{ provided } r+s = \frac{n}{p}, \tag{3.5.28}$$

assuming (3.5.26) holds. Of course, this also follows from the Sobolev imbedding theorem:

$$H^{r,p} \subset L^{np/(n-rp)}, \quad r < \frac{n}{p}, \tag{3.5.29}$$

together with Hölder's inequality, so it provides only a minor illustration of the effectiveness of Proposition 3.5.B.

As a more substantial illustration, we provide a proof of the following estimate, which for $p=2$ was announced in [Che2] and applied to interesting results on the Navier-Stokes equations.

PROPOSITION 3.5.C. *Let $u \in H^{s,p}$, $v \in H^{r,p}$ be vector valued. Assume $r < \frac{n}{p}$, $s < \frac{n}{p}+1$, $r+s \geq \frac{n}{p}$. Then*

$$\textit{div } v = 0 \Longrightarrow v \cdot \nabla u \in H^{r+s-\frac{n}{p}-1,p}. \tag{3.5.30}$$

PROOF: Using $v \cdot \nabla u = \text{div }(u \otimes v) - u(\text{div } v)$, we can write

$$v \cdot \nabla u = T_v(\nabla u) + T_{(\nabla u)} v + \text{ div } R(u,v). \tag{3.5.31}$$

The hypotheses on u and v imply

$$T_v \in OP\mathcal{B}^0 S^{\frac{n}{p}-r}_{1,1}, \quad T_{\nabla u} \in OP\mathcal{B}^0 S^{\frac{n}{p}-s+1}_{1,1},$$

so the first two terms on the right side of (3.5.31) belong to the space in (3.5.30). Whenever $r < n/p$ and $r+s > n/p$, we can apply (3.5.11) to obtain

$$R(u,v) \in H^{r+s-\frac{n}{p},p}. \tag{3.5.32}$$

It remains to consider the case $r < n/p$ and $r+s = n/p$. Then we can apply Proposition 3.5.B, with $f = u \in X^{-r} = H^{\frac{n}{p}-r,p}$, $g = v \in H^{r,p}$, to get $R(u,v) \in L^p$ in this case, so again div $R(u,v)$ belongs to the space in (3.5.31).

We record a further extension of Proposition 3.5.B.

PROPOSITION 3.5.D. *Given $r \in \mathbb{R}$, let X^r have the property* (3.5.21). *Then, for any $s \in [0,\infty)$, $p \in (1,\infty)$,*

$$\|R(f,g)\|_{H^{s,p}} \leq C\|f\|_{X^r}\|g\|_{H^{s-r,p}}. \tag{3.5.33}$$

PROOF: The content of (3.5.22) is that

$$R_f : H^{-r,p} \longrightarrow L^p \text{ for } f \in X^r.$$

We claim $R_f : H^{1-r,p} \to H^{1,p}$ for $f \in X^r$. This follows from

$$\partial_j R_f g = R_{\partial_j f} g + R_f \partial_j g.$$

By induction, one has $R_f : H^{j-r,p} \to H^{j,p}$ for $j = 1,2,3,\ldots$, when $f \in X^r$. Then, by interpolation, $R_f : H^{s-r,p} \to H^{s,p}$, $s \geq 0$, giving (3.5.33).

Above, we made use of the extension of (3.5.18) to L^p-estimates, based on $R_f \in OPS^0_{1,1}$. In fact, it is useful to note that, in general, (3.5.16) can be extended to L^p estimates. This is based on the following simple estimate.

LEMMA 3.5.E. *If $\varphi, \psi \in \mathcal{S}(\mathbb{R}^n)$, and $\tau_a f = \tau(a,f)$ is given by* (3.5.17), *then*

$$\tau_a f(x) = \int K_a(x,y) f(y)\,dy \tag{3.5.34}$$

with

$$\begin{aligned} |K_a(x,y)| &\leq C\|a\|_{C^0_*}|x-y|^{-n}, \\ |\nabla_{x,y} K_a(x,y)| &\leq C\|a\|_{C^0_*}|x-y|^{-n-1}. \end{aligned} \tag{3.5.35}$$

PROOF: We have the formula

$$K_a(x,y) = \sum m_k a_k(x) 2^{nk} \hat{\varphi}(2^k(x-y))$$

with $A_k = \psi(2^{-k}D)a$, so $\|a_k\|_{L^\infty} \leq C\|a\|_{C^0_*}$. From this the estimates (3.5.35) are easily obtained.

As discussed in §0.11, it is a basic result of Calderon and Zygmund that, when an L^2-bounded operator has a kernel satisfying (3.5.35), then the operator is of weak type $(1,1)$, and bounded on L^p for $p \in (1,\infty)$. Thus the estimate (3.5.16) extends to

$$\|\tau(a,f)\|_{L^p} \le C_p \|a\|_{BMO} \|f\|_{L^p}, \quad 1 < p < \infty. \tag{3.5.36}$$

Propositions 3.5.B and 3.5.D cannot extend completely to general $\tau(a,f)$, since in particular they cannot extend completely for $T_f a$, but the following partial extension is very useful.

PROPOSITION 3.5.F. *Let $\varphi, \psi \in \mathcal{S}(\mathbb{R}^n)$ and assume $\psi(\xi) = 0$ for $|\xi| \le c_0$. Let $\tau(a,f)$ be given by (3.5.17). Let $r \in \mathbb{Z}^+$ and assume X^r satisfies (3.5.21). Then*

$$\|\tau(a,f)\|_{H^{s,p}} \le C \|a\|_{X^r} \|f\|_{H^{s-r,p}}, \quad 0 \le s \le r. \tag{3.5.37}$$

PROOF: First take the case $s = 0$. Say $f = D^\alpha g$, $|\alpha| = r$, $g \in L^p$. Write $a = \sum_{\ell=1}^{10} a_\ell$, with $\hat{a}_\ell$ supported in $2^k \le |\xi| \le 2^{k+1}$ with $k = \ell \bmod 10$, such that there exist $b_\ell \in BMO$ with

$$\psi(2^{-k}D) a_\ell = 2^{-kr} \psi(2^{-k}D) b_\ell, \quad \forall k \in \mathbb{Z}^+,$$

as in the proof of Proposition 3.5.B. Thus, denoting (3.5.17) by $\tau_{\psi,\varphi}(a,f)$, we have

$$\tau_{\psi,\varphi}(a_\ell, D^\alpha g) = \tau_{\psi,\varphi_\alpha}(b_\ell, g) \tag{3.5.38}$$

where $\varphi_\alpha(\xi) = \xi^\alpha \varphi(\xi)$. Thus (3.5.36) yields (3.5.37) for $s = 0$. For $s = j \le r$, $f \in H^{j-r,p}$, write

$$D^\beta \tau(a,f) = \sum_{\gamma+\sigma=\beta} C_{\gamma\sigma} \tau(D^\gamma a, D^\sigma f), \tag{3.5.39}$$

and use induction to obtain (3.5.37) for such integers s. The general result for $0 \le s \le r$ follows by interpolation.

In particular this applies to $\tau(a,f) = T_f a$, so for example we have the estimate

$$\|T_f g\|_{H^{s,p}} \le C \|f\|_{H^{s-1,p}} \|g\|_{Lip^1}, \quad 0 \le s \le 1, \tag{3.5.40}$$

which will be useful for commutator estimates in the next section.

To end this section, we note results on the terms in the decomposition of

$$M(u;x,D) = M^\#(u;x,D) + M^b(u;x,D) \tag{3.5.41}$$

for $M(u;x,D)$ arising from $F(x,D^m u)$ as in (3.3.6), so

$$F(x,D^m u) = M(u;x,D)u \bmod C^\infty, \tag{3.5.42}$$

when the Bony-Meyer paradifferential method is used to construct $M^\#$, so

$$u \in C^{m+r} \Longrightarrow M^\#(u;x,\xi) \in \mathcal{B}^r S^m_{1,1} \quad M^b(u;x,\xi) \in S^{m-r}_{1,1}. \tag{3.5.43}$$

Estimates of the form above for $T_w v$ apply to $M^\#(u;x,D)$ and estimates of the form above for $R(w,v)$ apply to $M^b(u;x,D)$. Looking at the analysis behind (3.5.5), (3.5.33), (3.5.37), and (3.5.40), one verifies the following. The spaces X^r are as in Propositions 3.5.B and 3.5.D.

PROPOSITION 3.5.G. *For the decomposition* (3.5.41) *described above for* $M(u;x,D)$, *arising as in Proposition* 3.3.A, *we have the following estimates, for* $p \in (1,\infty)$:

$$\|M^\#(u;x,D)v\|_{H^{s,p}} \le C_{sp}(\|u\|_{L^\infty})\|u\|_{C^m}\|v\|_{H^{s+m,p}}, \quad s \in \mathbb{R}, \tag{3.5.44}$$

and, for $r \in \mathbb{R}$,

$$\|M^b(u;x,D)v\|_{H^{s,p}} \le C_{sp}(\|u\|_{L^\infty})\|u\|_{X^{m+r}}\|v\|_{H^{s+m-r,p}}, \quad s \ge 0. \tag{3.5.45}$$

Furthermore, given $r \in \mathbb{Z}^+$,

$$\|M^\#(u;x,D)v\|_{H^{s,p}} \le C(\|u\|_{L^\infty})\|v\|_{X^{m+r}}\|u\|_{H^{s+m-r,p}}, \quad 0 \le s \le r; \tag{3.5.46}$$

in particular,

$$\|M^\#(u;x,D)v\|_{H^{s,p}} \le C(\|u\|_{L^\infty})\|v\|_{C^{m+1}}\|u\|_{H^{s+m-1,p}}, \quad 0 \le s \le 1. \tag{3.5.47}$$

§3.6. Commutator estimates

In this section we establish a number of estimates, including the following two:

$$\|P(fu) - fPu\|_{L^p} \le C\|f\|_{Lip^1}\|u\|_{H^{s-1,p}} + C\|f\|_{H^{s,p}}\|u\|_{L^\infty} \tag{3.6.1}$$

given $P \in OPS^s_{1,0}$, $s > 0$, $p \in (1,\infty)$, and

$$\|P(fu) - fPu\|_{L^p} \le C\|f\|_{Lip^1}\|u\|_{L^p} \tag{3.6.2}$$

given $P \in OPS^1_{1,0}$. The first generalizes an estimate of Moser [Mo] when P is a differential operator. Such an estimate was proved by Kato and

Ponce [KP] when $P = (1-\Delta)^{s/2}$, by a different method than used here. The second estimate is due to Coifman-Meyer [CM2], generalizing the case when $P \in OPS^1_{cl}$ due to Calderon [Cal]. Both these estimates will play important roles in subsequent chapters.

To begin, write, as in (3.5.1),

$$\begin{aligned} f(Pu) &= T_f Pu + T_{Pu} f + R(f, Pu) \\ P(fu) &= PT_f u + PT_u f + PR(f, u). \end{aligned} \tag{3.6.3}$$

Using Theorem 3.4.A and the observations leading to Corollary 3.4.G, we have

$$f \in Lip^1, \ P \in OPS^s_{1,0} \Longrightarrow [T_f, P] \in OP\mathcal{B}^0 S^{s-1}_{1,1}. \tag{3.6.4}$$

hence

$$\|[T_f, P]u\|_{H^{\sigma,p}} \leq C\|f\|_{Lip^1} \|u\|_{H^{s-1+\sigma,p}}, \quad \sigma \in \mathbb{R}, \tag{3.6.5}$$

for $s \in \mathbb{R}, \ p \in (1, \infty)$.

We estimate separately the other four terms on the right sides in (3.6.3). First,

$$u \in L^\infty \Longrightarrow T_u \in OP\mathcal{B}^0 S^0_{1,1}, \tag{3.6.6}$$

so

$$\|PT_u f\|_{H^{\sigma,p}} \leq C\|u\|_{L^\infty} \|f\|_{H^{s+\sigma,p}}, \quad \sigma \in \mathbb{R}, \tag{3.6.7}$$

for $s \in \mathbb{R}, \ p \in (1, \infty)$. Similarly, by (3.5.7),

$$u \in L^\infty, \ P \in OPS^s_{1,0} \Longrightarrow T_{Pu} \in OP\mathcal{B}^0 S^s_{1,0} \text{ if } s > 0, \tag{3.6.8}$$

so

$$\|T_{Pu} f\|_{H^{\sigma,p}} \leq C\|u\|_{L^\infty} \|f\|_{H^{s+\sigma,p}}, \quad \sigma \in \mathbb{R}, \tag{3.6.9}$$

for $s > 0, \ p \in (1, \infty)$.

To estimate $R(f, Pu)$, we can use

$$f \in \text{Lip}^1 \Longrightarrow R_f \in OPS^{-1}_{1,1} \tag{3.6.10}$$

to deduce

$$\|R(f, Pu)\|_{H^{\sigma,p}} \leq C\|f\|_{Lip^1} \|u\|_{H^{s-1+\sigma,p}}, \quad \sigma > 0. \tag{3.6.11}$$

But since we are particularly interested in the case $\sigma = 0$ in (3.6.1), we will appeal to Proposition 3.5.D, with $X^r = X^1 = Lip^1$, to obtain

$$\|R(f, Pu)\|_{H^{\sigma,p}} \leq C\|f\|_{Lip^1}\|u\|_{H^{s-1+\sigma,p}}, \quad \sigma \geq 0. \tag{3.6.12}$$

Similarly, using this proposition, we have

$$\|PR(f, u)\|_{H^{\sigma,p}} \leq C\|f\|_{Lip^1}\|u\|_{H^{s-1+\sigma,p}}, \quad \sigma \geq -s, \tag{3.6.13}$$

such an estimate for $\sigma > -s$ following more simply from (3.6.10).

Thus the estimates (3.6.5), (3.6.7), (3.6.9), (3.6.12), and (3.6.13) yield the following extension of the Kato-Ponce estimate.

PROPOSITION 3.6.A. *Given $P \in OPS^s_{1,0}$, $s > 0$, we have*

$$\|P(fu) - fPu\|_{H^{\sigma,p}} \leq C\|f\|_{Lip^1}\|u\|_{H^{s-1+\sigma,p}} + C\|f\|_{H^{s+\sigma,p}}\|u\|_{L^\infty}, \tag{3.6.14}$$

for $\sigma \geq 0$, $p \in (1, \infty)$.

The one point in the proof of this proposition which depends on Theorem 3.5.A, i.e., Theorem 33 of [CM], is the estimate (3.6.12), improving (3.6.11). Thus (3.6.14) is proved for $\sigma > 0$ without using this result of [CM]. We now show how it can also be proved for $\sigma = 0$, independently of Theorem 3.5.A, using some simple commutations.

The only element of (3.6.3) that did not give rise to an adequate estimate for $\sigma = 0$, without depending on Theorem 3.5.A, is $R(f, Pu)$. However, we will find it more convenient to estimate

$$T^R_{Pu}f = T_{Pu}f + R(f, Pu), \tag{3.6.15}$$

where generally

$$fg = T_f g + T^R_g f; \quad T^R_g f = T_g f + R(f, g). \tag{3.6.16}$$

The advantage is that, from (3.6.6) and the obvious estimate on fg, we have

$$\|T^R_g f\|_{L^p} \leq C\|f\|_{L^\infty}\|g\|_{L^p}, \tag{3.6.17}$$

without using Theorem 3.5.A. We need to prove

$$\|T^R_{Pu}f\|_{L^p} \leq C\|u\|_{L^\infty}\|f\|_{H^{s,p}} + C\|u\|_{H^{s-1,p}}\|f\|_{Lip^1}. \tag{3.6.18}$$

Given $u \in H^{s-1,p} \cap L^\infty$, we can write

$$u = \sum \Lambda^{-1}\partial_j v_j + \Lambda^{-1}v_0, \quad \Lambda^s = (1 - \Delta)^{s/2}, \tag{3.6.19}$$

with $\|v_j\|_{H^{s-1,p}} \le C\|u\|_{H^{s-1,p}}$, $\|v_j\|_{C^0_*} \le C\|u\|_{L^\infty}$. It suffices to examine the case $u = \Lambda^{-1}\partial_j v_j$. Write

$$P_j = [P\Lambda^{-1}, \partial_j] \in OPS^{s-1}_{1,0}, \quad Q = P\Lambda^{-1} \in OPS^{s-1}_{1,0}.$$

Now, with $u = \Lambda^{-1}\partial_j v_j$, since $T^R_{\partial_j w} f = \partial_j T^R_w f - T^R_w \partial_j f$, we have

$$T^R_{Pu} f = T^R_{P_j v_j} f + \partial_j T^R_{Qv_j} f - T^R_{Qv_j}\partial_j f. \tag{3.6.20}$$

We estimate the three terms on the right. By (3.6.17),

$$\begin{aligned} \|T^R_{Qv_j}\partial_j f\|_{L^p} &\le C\|\partial_j f\|_{L^\infty}\|Qv_j\|_{L^p} \\ &\le C\|f\|_{Lip^1}\|v_j\|_{H^{s-1,p}}, \end{aligned} \tag{3.6.21}$$

and $T^R_{P_j v_j} f$ has an even simpler bound. To estimate the other term, i.e., the $H^{1,p}$-norm of $T^R_{Qv_j} f$, we will use the fact that

$$v \in C^0_* \Longrightarrow T^R_{Qv} \in OPS^{s-1}_{1,1}, \text{ if } s > 1, \tag{3.6.22}$$

to get

$$\|\partial_j T^R_{Qv_j} f\|_{L^p} \le C\|v_j\|_{C^0_*}\|f\|_{H^{s,p}}, \text{ if } s > 1. \tag{3.6.23}$$

Consequently, we have a proof of (3.6.1) which does not depend on Theorem 3.5.A, as long as $s > 1$.

We now turn to the commutator estimate (3.6.2), which is sharper than the $s = 1$ case of (3.6.1). From what we have done so far, we see from (3.6.5), (3.6.12), and (3.6.13) that, for $P \in OPS^1_{1,0}, \sigma \ge 0$,

$$\begin{aligned} &\|P(fu) - fPu\|_{H^{\sigma,p}} \\ &\quad \le C\|f\|_{Lip^1}\|u\|_{H^{\sigma,p}} + C\|PT_u f\|_{H^{\sigma,p}} + C\|T_{Pu} f\|_{H^{\sigma,p}}. \end{aligned} \tag{3.6.24}$$

Before, we dominated the last two terms by $C\|u\|_{L^\infty}\|f\|_{H^{s,p}}$, using (3.6.6) and (3.6.8). This time, we use instead (3.5.40), i.e.,

$$\|T_v f\|_{H^{r,p}} \le C\|v\|_{H^{r-1,p}}\|f\|_{Lip^1}, \quad 0 \le r \le 1, \tag{3.6.25}$$

to dominate the last two terms in (3.6.24) by $C\|f\|_{Lip^1}\|u\|_{L^p}$, in case $\sigma = 0$. This proves (3.6.2).

We mention that another path from Theorem 3.5.A to the commutator estimate (3.6.2) is indicated in Chapter 6 of [CM].

We briefly discuss how the estimate (3.6.2) follows from the T(1) Theorem of David-Journe [DJ]. First we recall the statement of that result. Consider a function K on $\mathbb{R}^n \times \mathbb{R}^n$ satisfying

$$
\begin{aligned}
|K(x,y)| &\le C|x-y|^{-n}, \\
|\nabla_{x,y} K(x,y)| &\le C|x-y|^{-n-1}.
\end{aligned}
\tag{3.6.26}
$$

Suppose K agrees on $\mathbb{R}^n \times \mathbb{R}^n \setminus \Delta$ with the distributional kernel of an operator $T : C_0^\infty \to \mathcal{D}'$, satisfying the *weak boundedness condition*

$$
|\langle T\varphi^{y,\lambda}, \psi^{y,\lambda}\rangle| \le C\lambda^n \tag{3.6.27}
$$

for all φ, ψ in any bounded subset of $C_0^\infty(\mathbb{R}^n)$, where

$$
\varphi^{y,\lambda}(x) = \varphi\big((x-y)/\lambda\big).
$$

Then the Theorem states that there is a bounded extension

$$
T : L^p(\mathbb{R}^n) \longrightarrow L^p(\mathbb{R}^n), \quad 1 < p < \infty, \tag{3.6.28}
$$

provided that also

$$
T(1),\ T^*(1) \in BMO. \tag{3.6.29}
$$

We recall the following property, which actually characterizes BMO :

$$
A \in OPS^0_{1,0} \Longrightarrow A : L^\infty \longrightarrow BMO. \tag{3.6.30}
$$

We claim that, if $P \in OPS^1_{1,0}$ and $f \in \text{Lip}^1$, then $T = [P, f]$ has these properties. If P has Schwartz kernel $L(x, x-y)$, then the Schwartz kernel of T is

$$
K(x,y) = L(x, x-y)[f(x) - f(y)], \tag{3.6.31}
$$

which implies (3.6.26) from standard estimates on L. Note that

$$
\begin{aligned}
T(1) &= P(f) - fP(1) \in BMO, \\
T^*(1) &= -P^*(\overline{f}) + \overline{f}P^*(1) \in BMO,
\end{aligned}
\tag{3.6.32}
$$

by (3.6.30). It remains to verify the weak boundedness condition (3.6.27).

When $P \in OPS^1_{1,0}$ has *symmetric* Schwartz kernel, then $T = [P, f]$ has *antisymmetric* Schwartz kernel. As noted in [DJ], the estimate (3.6.27) is easy in this case. Furthermore, it is easy to see that $P \in OPS^1_{1,0}$ has a symmetric Schwartz kernel (mod $OPS^0_{1,0}$) if and only if its symbol $p(x, \xi) \in$

$S^1_{1,0}$ is symmetric in ξ (mod $S^0_{1,0}$). There is a corresponding result on antisymmetry.

Thus it remains to consider the case of $p(x,\xi) \in S^1_{1,0}$ antisymmetric in ξ. Hence we can consider $P = \sum D_j B_j(x,D)$, with $B_j(x,\xi) \in S^0_{1,0}$, symmetric in ξ. Now we have

$$[D_j B_j(x,D), f] = [D_j, f]B_j(x,D) + D_j[B_j(x,D), f]. \tag{3.6.33}$$

The first term on the right is obviously bounded on L^p. Write the second as $(D_j\Lambda^{-1})\Lambda[B_j(x,D), f]$, and write

$$\Lambda[B_j(x,D), f] = [\Lambda B_j(x,D), f] - [\Lambda, f]B_j(x,D). \tag{3.6.34}$$

The operators $\Lambda B_j(x,D)$, $\Lambda \in OPS^1_{1,0}$ both have even kernels (mod $OPS^0_{1,0}$), so the case discussed above applies. Thus we have the L^p-boundedness of (3.6.33), finishing the derivation of (3.6.2) from the $T(1)$ Theorem.

The $T(1)$ Theorem is proved in [DJ]. A discussion of background material is given in [Ch]. We note that a key ingredient in the proof is the paraproduct, including Theorem 3.5.A.

To close this section, we extend Proposition 3.6.A to the case $s = 0$. In the derivation of (3.6.14), only the step involving (3.6.9) required $s > 0$. We can extend this result to $s = 0$ if we replace $\|u\|_{L^\infty}$ on the right side of (3.6.14) by $\|u\|_{L^\infty} + \|Pu\|_{L^\infty}$. We also note that, using (3.6.25), we can estimate $T_{Pu}f$ and PT_uf in the $H^{\sigma,p}$-norm by $C\|f\|_{Lip^1}\|u\|_{H^{\sigma-1,p}}$, for $0 \le \sigma \le 1$. Thus we have the following.

PROPOSITION 3.6.B. *Given $P \in OPS^0_{1,0}$, we have*

$$\|P(fu) - fPu\|_{H^{\sigma,p}} \le C\|f\|_{Lip^1}\|u\|_{H^{\sigma-1,p}}, \quad \text{for } 0 \le \sigma \le 1, \tag{3.6.35}$$

and, for $\sigma > 1$,

$$\begin{aligned} &\|P(fu) - fPu\|_{H^{\sigma,p}} \\ &\quad \le C\|f\|_{Lip^1}\|u\|_{H^{\sigma-1,p}} + C\|f\|_{H^{\sigma,p}}\big(\|u\|_{L^\infty} + \|Pu\|_{L^\infty}\big). \end{aligned} \tag{3.6.36}$$

Chapter 4: Calculus for $OPC^1S^m_{cl}$

In the last chapter, we developed an operator calculus and used it for several purposes, including obtaining commutator estimates in §3.6. Here we work in the opposite order. In §4.1 we recall the estimate (3.6.2) of Coifman-Meyer (generalizing results of Calderon) and show how it leads to further commutator estimates for operators with C^1-regular symbols. Then we use these commutator estimates to establish an operator calculus for symbols in $C^1S^m_{cl}$. For this, Calderon's estimates suffice, and much of the material of §4.2 is contained in [Ca2], [Ca3]. In §4.3, we look at a Gårding inequality, more precise, though less general, than the Gårding inequality in Proposition 2.4.B.

Section 4.4 discusses relations between $OPC^1S^m_{cl}$ and paradifferential operators with symbols in Bony's class Σ^m_1.

§4.1. Commutator estimates

We begin by recalling the commutator estimate of Coifman-Meyer proved in §3.6.

PROPOSITION 4.1.A. *If* $P \in OPS^1_{1,0}$, $p \in (1,\infty)$, *then*

$$\|P(fu) - fPu\|_{L^p} \le C\|f\|_{Lip^1}\|u\|_{L^p}. \tag{4.1.1}$$

We will derive several generalizations of this, with implications for $OPC^1S^m_{cl}$. To begin, we obtain a generalization of (4.1.1) with f replaced by $A(x,D) \in OPC^1S^0_{cl}$, as follows. Modulo minor additional terms,

$$A(x,D) = \sum_\ell a_\ell(x)\omega_\ell(D), \tag{4.1.2}$$

as noted in the proof of Proposition 1.1.A. We have

$$\|a_\ell\|_{C^1} \le C_M\langle\ell\rangle^{-M}, \tag{4.1.3}$$

where C_M is dominated by some seminorm

$$\pi^0_{N,C^1}(A) = \sup\{\|D^\alpha_\xi A(\cdot,\xi)\|_{C^1}\cdot\langle\xi\rangle^{|\alpha|} : \xi \in \mathbb{R}^n, |\alpha| \le N\}. \tag{4.1.4}$$

Meanwhile we have a polynomial bound

$$\|\omega_\ell(D)\|_{\mathcal{L}(L^p)} \le C_p\langle\ell\rangle^K, \quad 1 < p < \infty. \tag{4.1.5}$$

Given (4.1.2), we have

$$[P, A(x,D)] = \sum_\ell [P, a_\ell]\omega_\ell(D) + \sum_\ell a_\ell(x)[P,\omega_\ell(D)]. \tag{4.1.6}$$

We can apply (4.1.1) to $[P, a_\ell]$. To analyze the last sum in (4.1.6), we use the estimate

$$\big\|[P, \omega_\ell(D)]\big\|_{\mathcal{L}(L^p)} \leq C_p \langle \ell \rangle^K, \tag{4.1.7}$$

coming from $[P, \omega_\ell(D)] \in OPS^0_{1,0}$, with easily established symbol bounds. Putting these estimates together, we have:

PROPOSITION 4.1.B. *If $P \in OPS^1_{1,0}$, $A(x,\xi) \in C^1 S^0_{cl}$, $1 < p < \infty$, then, for some N,*

$$\big\|[P, A(x,D)]u\big\|_{L^p} \leq C_p \pi^0_{N,C^1}(A) \|u\|_{L^p}. \tag{4.1.8}$$

In the same way, given $B(x,\xi) \in C^1 S^1_{cl}$, we can write

$$B(x,D) = \sum_\ell b_\ell(x) \omega_\ell(D) \Lambda \tag{4.1.9}$$

and then

$$[B(x,D), A(x,D)] = \sum_{\ell,m} [b_m(x)\omega_m(D)\Lambda, a_\ell(x)\omega_\ell(D)]. \tag{4.1.10}$$

Each commutator in the sum can be expanded to

$$b_m[\omega_m(D)\Lambda, a_\ell]\omega_\ell + a_\ell[b_m\Lambda, \omega_\ell]\omega_m. \tag{4.1.11}$$

We can apply (4.1.1) to the first term in (4.1.11). We can rewrite the second term, using

$$[b_m\Lambda, \omega_\ell] = [b_m, \omega_\ell \Lambda] - \omega_\ell[b_m, \Lambda],$$

and again apply (4.1.1). Thus we have:

PROPOSITION 4.1.C. *If $B(x,\xi) \in C^1 S^1_{cl}$, $A(x,\xi) \in C^1 S^0_{cl}$, $1 < p < \infty$, then, for some N,*

$$\|[B(x,D), A(x,D)]u\|_{L^p} \leq C \pi^0_{N,C^1}(A) \cdot \pi^1_{N,C^1}(B) \|u\|_{L^p}. \tag{4.1.12}$$

Here, generalizing (4.1.4), we set

$$\pi^s_{N,X}(B) = \sup\big\{\|D^\alpha_\xi B(\cdot,\xi)\|_X \cdot \langle \xi \rangle^{-s+|\alpha|} : \xi \in \mathbb{R}^n, |\alpha| \leq N\big\}. \tag{4.1.13}$$

Note that, under the hypotheses above, by Proposition 1.1.B,

$$\begin{aligned} A(x,D)B(x,D) &: H^{s+1,p} \longrightarrow H^{s,p}, \ -1 \le s \le 1 \\ B(x,D)A(x,D) &: H^{s+1,p} \longrightarrow H^{s,p}, \quad 0 \le s \le 1, \end{aligned} \tag{4.1.14}$$

for $1 < p < \infty$.

Noting that $\Lambda[P,f] = [\Lambda P, f] - [\Lambda, f]P$ and $[P,f]\Lambda = [P\Lambda, f] - P[\Lambda, f]$, we see that Proposition 4.1.A implies:

PROPOSITION 4.1.D. *If $P \in OPS^0_{1,0}$, $f \in Lip^1$, $1 < p < \infty$, then*

$$[P,f] : H^{s,p} \longrightarrow H^{s+1,p}, \quad -1 \le s \le 0.$$

Then the analysis establishing Proposition 4.1.B and Proposition 4.1.C also gives:

PROPOSITION 4.1.E. *If $P \in OPS^0_{1,0}$, $A_j(x,\xi) \in C^1S^0_{cl}$, $1 < p < \infty$, then*

$$[P, A_j(x,D)] : H^{s,p} \longrightarrow H^{s+1,p}, \quad -1 \le s \le 0$$

and

$$[A_1(x,D), A_2(x,D)] : H^{s,p} \longrightarrow H^{s+1,p}, \quad -1 \le s \le 0.$$

We end this section with a generalization of the Kato-Ponce estimate (3.6.1), replacing $f \in \text{Lip}^1 \cap H^{s,p}$ by $A(x,D) \in OPC^1S^0_{cl} \cap OPH^{s,p}S^0_{cl}$, given $1 < p < \infty, s > 0$. Write $A(x,D)$ in the form (4.1.2). Then, given $P \in OPS^s_{1,0}$,

$$[P, A(x,D)]u = \sum_\ell [P, a_\ell]\omega_\ell(D)u + \sum_\ell a_\ell(x)[P, \omega_\ell(D)]u. \tag{4.1.15}$$

By (3.5.1), with $f = a_\ell$, we have

(4.1.16)
$$\big\|[P,a_\ell](\omega_\ell(D)u)\big\|_{L^p} \le C\|a_\ell\|_{C^1}\|\omega_\ell(D)u\|_{H^{s-1,p}} + C\|a_\ell\|_{H^{s,p}}\|\omega_\ell(D)u\|_{L^\infty}.$$

Now a_ℓ has a bound of the form (4.1.3), both in the C^1-norm and in the $H^{s,p}$-norm. In light of (4.1.5), we have

$$C\|a_\ell\|_{C^1}\|\omega_\ell(D)u\|_{H^{s-1,p}} \le C_{pM}\langle \ell\rangle^{-M+K}\pi^0_{N,C^1}(A)\|u\|_{H^{s-1,p}}. \tag{4.1.17}$$

Now OPS^0 is not bounded on L^∞, so we need to work with a smaller Banach space, call it $C^0_\#$, with the property that

$$B \in OPS^0_{cl} \Longrightarrow B : C^0_\# \longrightarrow C^0 \subset L^\infty. \tag{4.1.18}$$

Then we have an estimate

$$C\|a_\ell\|_{H^{s,p}}\|\omega_\ell(D)u\|_{L^\infty} \le C_{pM}\langle\ell\rangle^{-M+K}\pi^0_{N,H^{s,p}}(A)\|u\|_{C^0_\#}. \tag{4.1.19}$$

We have then a good estimate on the first sum in (4.1.15). For the last sum, we can use the simple estimate

$$\big\|[P,\omega_\ell(D)]u\big\|_{L^p} \le C_p\langle\ell\rangle^K\|u\|_{H^{s-1,p}}.$$

We have the following result:

PROPOSITION 4.1.F. *If* $P \in OPS^s_{1,0}$ *and* $A(x,\xi) \in C^1S^0_{cl} \cap H^{s,p}S^0_{cl}$ *with* $1 < p < \infty, s > 0$, *then*

$$\|[P, A(x,D)]u\|_{L^p} \le C\pi^0_{N,C^1}(A)\|u\|_{H^{s-1,p}} + C\pi^0_{N,H^{s,p}}(A)\|u\|_{C^0_\#}. \tag{4.1.20}$$

Whatever choice of $C^0_\#$ is made in (4.1.18), we note that, if

$$C^r_\# = \Lambda^{-r}(C^0_\#), \quad r \ge 0, \tag{4.1.21}$$

then

$$B \in OPS^0_{cl} \Longrightarrow B : C^r_\# \longrightarrow C^r, \quad r \ge 0. \tag{4.1.22}$$

For $r \notin \mathbb{Z}$, this is clear from $C^0 \subset C^0_*$ and $\Lambda^{-r} : C^0_* \longrightarrow C^r_*$. If $r = k \in \mathbb{Z}$, $u = \Lambda^{-r}v$, $v \in C^0_\#$, we have, for $|\alpha| \le k$, $D^\alpha B\Lambda^{-k}v = Av$, $A \in OPS^0_{cl}$, which belongs to C^0 by hypothesis (4.1.18), so (4.1.22) is established. One candidate for $C^0_\#$ is the Besov-type space

$$B^0_{\infty,1} = \{u \in \mathcal{S}'(\mathbb{R}^n) : \sum_{j\ge 0}\|\psi_j(D)u\|_{L^\infty} < \infty\}, \tag{4.1.23}$$

where $\{\psi_j\}$ is the partition of unity (1.3.1). Note that

$$C^0 \supset B^0_{\infty,1} \supset C^r, \quad r > 0. \tag{4.1.24}$$

However, for the purpose of estimates we need in §5.3, it is useful to make a different choice of $C^0_\#$. We will say

$$u \in C^0_\# \Longleftrightarrow \sup_\ell \langle\ell\rangle^{-K}\|\omega_\ell(D)u\|_{C^0} < \infty \tag{4.1.25}$$

where K is a constant which is picked to be sufficiently large. For details, see §B.2.

§4.2. Operator algebra

The analysis proving Proposition 4.1.C can also be applied to products $A_j(x,D)B(x,D)$ where, with $j=0$ or 1,

$$A_j(x,\xi)\in C^1S^j_{cl},\quad B(x,\xi)\in C^1S^\mu_{cl}. \tag{4.2.1}$$

We suppose expansions such as (4.1.2) and (4.1.9) hold. Then, in analogy with (4.1.10), we have

$$\begin{aligned}A_0(x,D)B(x,D)&=\sum_{\ell,m}a_\ell(x)\omega_\ell(D)b_m(x)\omega_m(D)\Lambda^\mu\\&=C_0(x,D)+R_0\end{aligned} \tag{4.2.2}$$

where

$$C_0(x,\xi)=A_0(x,\xi)B(x,\xi) \tag{4.2.3}$$

and

$$R_0=\sum_{\ell,m}a_\ell(x)[\omega_\ell(D),b_m(x)]\Lambda^\mu\omega_m(D). \tag{4.2.4}$$

Applying Proposition 4.1.E, keeping in mind such estimates as (4.1.3), we have the first half of:

PROPOSITION 4.2.A. *Given* (4.2.1), $j=0,1$,

$$A_j(x,D)B(x,D)=C_j(x,D)+R_j, \tag{4.2.5}$$

where

$$C_j(x,\xi)=A_j(x,\xi)B(x,\xi)\in C^1S^{\mu+j}_{cl}, \tag{4.2.6}$$

$$R_0:H^{\mu+s,p}\longrightarrow H^{s+1,p},\quad -1\le s\le 0, \tag{4.2.7}$$

and

$$R_1:H^{\mu,p}\longrightarrow L^p. \tag{4.2.8}$$

For the second half, note that

$$R_1=\sum_{\ell,m}b_m(x)[\omega_m(D)\Lambda,a_\ell(x)]\omega_\ell(D)\Lambda^\mu. \tag{4.2.9}$$

Note that Proposition 4.2.A contains Proposition 4.1.C.

Next we establish a result on adjoints, more precise, though less general, than Proposition 2.3.A.

PROPOSITION 4.2.B. *If $B(x,\xi) \in C^1S^1_{cl}$, and $B^*(x,\xi) = B(x,\xi)^*$, then*

$$B(x,D)^* - B^*(x,D) : L^p \longrightarrow L^p, \quad 1 < p < \infty. \tag{4.2.10}$$

PROOF: Given the expansion (4.1.9), we have

$$B(x,D)^* - B^*(x,D) = \sum_\ell [\omega_\ell(D)\Lambda, b_\ell(x)^*], \tag{4.2.11}$$

so the boundedness follows from (4.1.1) and the analogue of (4.1.3).

In a similar fashion, we can use Proposition 4.1.D to establish the following.

PROPOSITION 4.2.C. *If $A(x,\xi) \in C^1S^0_{cl}$, then*

$$A(x,D)^* = A^*(x,D) + R \tag{4.2.12}$$

with

$$R : H^{s,p} \longrightarrow H^{s+1,p}, \quad -1 \le s \le 0. \tag{4.2.13}$$

§4.3. Gårding inequality

We use the operator algebra of §4.2 to establish the following useful version of Gårding's inequality.

PROPOSSITION 4.3.A. *If $p(x,\xi) \in C^1S^2_{cl}$ is a $K \times K$ matrix valued symbol and*

$$p(x,\xi) = p(x,\xi)^* \ge C|\xi|^2 I \tag{4.3.1}$$

for $|\xi|$ large, then

$$Re\ (p(x,D)u,u) \ge C_1\|u\|^2_{H^1} - C_2\|u\|^2_{L^2}. \tag{4.3.2}$$

PROOF: Without loss of generality, we can assume

$$p(x,\xi) = p(x,\xi)^* \ge C\langle\xi\rangle^2 I \text{ for all } x,\xi. \tag{4.3.3}$$

Picking $C_1 < C$, we can hence write

$$p(x,\xi) - C_1\langle\xi\rangle^2 = a(x,\xi)^2, \tag{4.3.4}$$

with $a(x,\xi) = a(x,\xi)^* \in C^1S^1_{cl}$. Hence

$$(p(x,D)u,u) = C_1\|u\|^2_{H^1} + (a^2(x,D)u,u) \tag{4.3.5}$$

where $a^2(x,\xi) = a(x,\xi)^2$. Applying Proposition 4.2.A and Proposition 4.2.B, we can write

$$a^2(x,D) = a(x,D)^*a(x,D) + R, \quad R : H^1 \longrightarrow L^2. \tag{4.3.6}$$

Hence

$$(p(x,D)u,u) = C_1\|u\|^2_{H^1} + \|a(x,D)u\|^2_{L^2} + (Ru,u) \tag{4.3.7}$$

and

$$|(Ru,u)| \leq C_3\|u\|_{H^1}\|u\|_{L^2}. \tag{4.3.8}$$

This gives the desired estimate (4.3.2).

We can replace (4.3.1) by a more standard hypothesis on $p(x,\xi) + p(x,\xi)^*$, though a little care is required to do so, since the relation between $p(x,D)^*$ and $p^*(x,D)$ is not as good for $p(x,\xi) \in C^1S^2_{cl}$ as it is for symbols in $C^1S^1_{cl}$.

LEMMA 4.3.B. *Given $q(x,\xi) = q(x,\xi)^* \in C^1S^2_{cl}$, we can write*

$$q(x,D) = Q + R \tag{4.3.9}$$

where Q is a symmetric operator on L^2, with domain H^2, and

$$R : H^1 \longrightarrow L^2. \tag{4.3.10}$$

PROOF: There exists $C_0 > 0$ such that $p(x,\xi) = q(x,\xi) + C_0\langle\xi\rangle^2$ satisfies (4.3.3). Then (4.3.4) and (4.3.6) give

$$q(x,D) = (C_1 - C_0)\Lambda + a(x,D)^*a(x,D) + R, \tag{4.3.11}$$

which does it. Note that it does not follow that $Q \in OPC^1S^2_{cl}$, nor that $q(x,D) - q(x,D)^*$ maps H^1 to L^2.

Consequently the hypothesis on $q(x,\xi)$ implies

$$|\text{Im}\,(q(x,D)u,u)| \leq C\|u\|_{H^1}\|u\|_{L^2}. \tag{4.3.12}$$

Therefore, if $p(x,\xi) \in C^1 S^2_{cl}$ and $p^s(x,\xi) = \frac{1}{2}(p(x,\xi) + p(x,\xi)^*)$, then

$$|\text{Re } (p(x,D)u,u) - \text{ Re } (p^s(x,D)u,u)| \le C\|u\|_{H^1}\|u\|_{L^2}. \tag{4.3.13}$$

This immediately yields

COROLLARY 4.3.C. *In Proposition 4.3.A, the hypothesis (4.3.1) can be replaced by*

$$p(x,\xi) + p(x,\xi)^* \ge C|\xi|^2 I. \tag{4.3.14}$$

§4.4. C^1-paradifferential calculus

In this section we discuss the relation of operators with symbols in $C^1 S^m_{cl}$ and those with symbols in the class Σ^m_1. This is the case of Bony's symbol class Σ^m_r with $r = 1$, which has played a role in recent work of Gerard-Rauch [GR] and Metivier [Met2]. Recall from (3.4.32) that a symbol in Σ^m_1 arises from one in $C^1 S^m_{cl}$ via a symbol smoothing process, with $\delta = 1$, of the form

$$a^\#(\cdot,\xi) = \sum_{k\ge 0} \Psi_{k-5}(D)a(\cdot,\xi)\psi_{k+1}(\xi). \tag{4.4.1}$$

As in §3.4, we follow Bony [Bo] and denote $a^\#(x,D)$ also by T_a. The first basic result is the following, part of which was stated by Metivier in (9.7) of [Met2].

PROPOSITION 4.4.A. *If $a(x) \in C^1$, then, for $-1 \le s \le 0,\ p \in (1,\infty)$,*

$$u \in H^{s,p} \Longrightarrow au - T_a u \in H^{s+1,p}. \tag{4.4.2}$$

PROOF: We write

$$au - T_a u = T_u a + R(a,u) \tag{4.4.3}$$

and use results of §3.5 to analyze the two terms on the right. From Proposition 3.5.D we have

$$\|R(a,u)\|_{H^{s+1,p}} \le C\|a\|_{C^1}\|u\|_{H^{s,p}}, \quad s \ge -1, \tag{4.4.4}$$

while (3.5.40) implies

$$\|T_u a\|_{H^{s+1,p}} \le C\|a\|_{C^1}\|u\|_{H^{s,p}}, \quad -1 \le s \le 0. \tag{4.4.5}$$

Thus (4.4.2) follows.

Using (4.4.2) it is fairly easy to establish:

PROPOSITION 4.4.B. *If* $a(x,\xi) \in C^1S^m_{cl}$, *then, for* $-1 \le s \le 0$, $p \in (1,\infty)$,

$$u \in H^{s+m,p} \Longrightarrow a(x,D)u - T_a u \in H^{s+1,p}. \tag{4.4.6}$$

PROOF: Writing $a(x,\xi) = \sum a_j(x)\beta_j(\xi)$ with $\beta_j \in S^m_{cl}$, it suffices to apply (4.4.2) to $a = a_j(x)$, with u replaced by $\beta_j(D)u$.

Using this we can relate such commutator estimates as in Proposition 4.1.E to that from Corollary 3.4.H, which implies for scalar $a_j(x,\xi) \in C^1S^{m_j}_{cl}$,

$$T_{a_1}T_{a_2} - T_{a_2}T_{a_1} \in \mathcal{L}(H^{s,p}, H^{s-m_1-m_2+1,p}) \tag{4.4.7}$$

for $s \in \mathbb{R}$, $p \in (1,\infty)$. Given Proposition 4.4.B, we see that Proposition 4.1.E is *equivalent* to (4.4.7), for $s \in [-1,0]$, and $m_1 = m_2 = 0$. Similarly, (4.4.7) implies (4.4.1) for $P \in OPS^1_{cl}$, which is Calderon's case of Proposition 4.1.A. Of course, we saw already in §3.6 that the full strength of Proposition 4.1.A follows from such ingredients, so this is nothing new.

Chapter 5: Nonlinear hyperbolic systems

In this chapter we treat various types of hyperbolic equations, beginning in §5.1 with first order symmetric hyperbolic systems. In this case, little direct use of pseudodifferential operator techniques is made, mainly an appeal to the Kato-Ponce estimates. We use Friedrichs mollifiers to set up a modified Galerkin method for producing solutions, and some of their properties, such as (5.1.43), can be approached from a pseudodifferential operator perspective. The idea to use Moser type estimates and to aim for results on persistence of solutions as long as the C^1-norms remain bounded was influenced by [Mj]. We provide a slight sharpening, demonstrating persistence of solutions as long as the C^1_*-norm is bounded. In §5.2 we study two types of symmetrizable systems, the latter type involving pseudodifferential operators in an essential way. Here and in subsequent sections, including a treatment of higher order hyperbolic equations, we make strong use of the $C^1S^m_{cl}$-calculus developed in Chapter 4.

§5.1. Quasilinear symmetric hyperbolic systems

In this section we examine existence, uniqueness, and regularity for solutions to a system of equations of the form

$$\frac{\partial u}{\partial t} = L(t,x,u,D_x)u + g(t,x,u), \quad u(0) = f. \tag{5.1.1}$$

We derive a short time existence theorem, under the following assumptions. We suppose

$$L(t,x,u,D_x)v = \sum_j A_j(t,x,u)\partial_j v, \tag{5.1.2}$$

that each A_j is a $K \times K$ matrix, smooth in its arguments, and furthermore symmetric:

$$A_j = A_j^*. \tag{5.1.3}$$

We suppose g is smooth in its arguments, with values in $\mathbb{R}^K$; $u = u(t,x)$ takes values in $\mathbb{R}^K$. We then say (5.1.1) is a symmetric hyperbolic system. For simplicity we will suppose $x \in M$ where M is an n-dimensional torus, though any compact M could be treated with minor modifications, as could the case $M = \mathbb{R}^n$. We will suppose $f \in H^s(M),\ s > \frac{n}{2} + 1$.

Our strategy will be to obtain a solution to (5.1.1) as a limit of solutions u_ε to

$$\partial u_\varepsilon/\partial t = J_\varepsilon L_\varepsilon J_\varepsilon u_\varepsilon + g_\varepsilon, \quad u_\varepsilon(0) = f, \tag{5.1.4}$$

where

$$L_\varepsilon v = \sum_j A_j(t, x, J_\varepsilon u_\varepsilon)\partial_j v \tag{5.1.5}$$

and

$$g_\varepsilon = J_\varepsilon g(t, x, J_\varepsilon u_\varepsilon). \tag{5.1.6}$$

In (5.1.4), f might also be replaced by $J_\varepsilon f$, though this is not crucial. Here $\{J_\varepsilon : 0 < \varepsilon \le 1\}$ is a Friedrichs mollifier. For any $\varepsilon > 0$, (5.1.4) can be regarded as a (Banach space)ODE for u_ε, for which we know there is a unique solution, for t close to 0. Our task will be to show that the solution u_ε exists for t in an interval independent of $\varepsilon \in (0, 1]$, and has a limit as $\varepsilon \to 0$ solving (5.1.1).

To do this we estimate the H^s-norm of solutions to (5.1.4). We use the norm $\|u\|_{H^s} = \|\Lambda^s u\|_{L^2}$. We can arrange for Λ^s and J_ε to commute. We proceed to derive an estimate for

$$(d/dt)\|\Lambda^s u_\varepsilon(t)\|_{L^2}^2 = 2(\Lambda^s J_\varepsilon L_\varepsilon J_\varepsilon u_\varepsilon, \Lambda^s u_\varepsilon) + 2(\Lambda^s g_\varepsilon, \Lambda^s u_\varepsilon). \tag{5.1.7}$$

Write the first term as

$$2(L_\varepsilon \Lambda^s J_\varepsilon u_\varepsilon, \Lambda^s J_\varepsilon u_\varepsilon) + 2([\Lambda^s, L_\varepsilon] J_\varepsilon u_\varepsilon, \Lambda^s J_\varepsilon u_\varepsilon). \tag{5.1.8}$$

To estimate the first term of (5.1.8), use

$$(L_\varepsilon + L_\varepsilon^*) = -\sum_j [\partial_j A_j(t, x, J_\varepsilon u_\varepsilon)]v, \tag{5.1.9}$$

so

$$2(L_\varepsilon \Lambda^s J_\varepsilon u_\varepsilon, \Lambda^s J_\varepsilon u_\varepsilon) \le C(\|J_\varepsilon u_\varepsilon(t)\|_{C^1}) \cdot \|\Lambda^s J_\varepsilon u_\varepsilon\|_{L^2}^2. \tag{5.1.10}$$

Next consider

$$[\Lambda^s, L_\varepsilon]v = \sum_j [\Lambda^s(A_{j\varepsilon}\partial_j v) - A_{j\varepsilon}\Lambda^s(\partial_j v)], \tag{5.1.11}$$

where $A_{j\varepsilon} = A_j(t, x, J_\varepsilon u_\varepsilon)$. By the Kato-Ponce estimate (3.6.1), we have

$$\|[\Lambda^s, L_\varepsilon]v\|_{L^2} \le C \sum_j \Big[\|A_{j\varepsilon}\|_{H^s}\|\partial_j v\|_{L^\infty} + \|A_{j\varepsilon}\|_{C^1}\|\partial_j v\|_{H^{s-1}}\Big]. \tag{5.1.12}$$

Also, there is the Moser estimate

$$(5.1.13)\qquad \|A_j(t,x,w)\|_{H^s} \le C(\|w\|_{L^\infty})(1+\|w\|_{H^s}),$$

and a similar estimate on $\|g_\varepsilon\|_{H^s}$; compare (3.1.20). Using these estimates, we obtain from (5.1.7) that

$$(5.1.14)\qquad (d/dt)\|\Lambda^s u_\varepsilon(t)\|_{L^2}^2 \le C(\|J_\varepsilon u_\varepsilon(t)\|_{C^1})(1+\|J_\varepsilon u_\varepsilon(t)\|_{H^s}^2).$$

This puts us in a position to prove the following.

LEMMA 5.1.A. *Given $f \in H^s$, $s > \frac{n}{2}+1$, the solution to* (5.1.4) *exists for t in an interval $I = (-A, B)$, independent of ε, and satisfies an estimate*

$$(5.1.15)\qquad \|u_\varepsilon(t)\|_{H^s} \le K(t), \quad t \in I,$$

independent of $\varepsilon \in (0,1]$.

PROOF: Using the Sobolev imbedding theorem, we can dominate the right side of (5.1.14) by $E(\|u_\varepsilon(t)\|_{H^s}^2)$, so $\|u_\varepsilon(t)\|_{H^s}^2 = y(t)$ satisfies the differential inequality

$$(5.1.16)\qquad dy/dt \le E(y), \quad y(0) = \|f\|_{H^s}^2.$$

Gronwall's inequality yields a function $K(t)$, finite on some interval $[0, B)$, giving an upper bound for all $y(t)$ satisfying (5.1.16). Time-reversal gives such an upper bound on an interval $(-A, 0]$. This $I = (-A, B)$ and $K(t)$ work for (5.1.15).

We are now prepared to establish the following existence result.

THEOREM 5.1.B. *Provided* (5.1.1) *is symmetric hyperbolic and $f \in H^s(M)$, with $s > \frac{n}{2}+1$, there is a solution u, on an interval I about 0, with*

$$(5.1.17)\qquad u \in L^\infty(I, H^s(M)) \cap Lip(I, H^{s-1}(M)).$$

PROOF: Take the I above and shrink it slightly. The bounded family

$$u_\varepsilon \in C(I, H^s) \cap C^1(I, H^{s-1})$$

will have a weak limit point u satisfying (5.1.17). Furthermore, by Ascoli's theorem, there is a sequence

$$(5.1.18)\qquad u_{\varepsilon_\nu} \longrightarrow u \text{ in } C(I, H^{s-1}(M))$$

since the inclusion $H^s \subset H^{s-1}$ is compact. Also, by interpolation inequalities, $\{u_\varepsilon : 0 < \varepsilon \le 1\}$ is bounded in $C^\sigma(I, H^{s-\sigma}(M))$ for $0 < \varepsilon \le 1$, so since the inclusion $H^{s-\sigma} \subset C^1(M)$ is compact for small $\sigma > 0$ if $s > \frac{n}{2}+1$, we can arrange that

$$u_{\varepsilon_\nu} \longrightarrow u \text{ in } C(I, C^1(M)). \tag{5.1.19}$$

Consequently (with $\varepsilon = \varepsilon_\nu$)

$$\begin{aligned} &J_\varepsilon L(t,x,J_\varepsilon u_\varepsilon, D)J_\varepsilon u_\varepsilon + J_\varepsilon g(t,x,J_\varepsilon u_\varepsilon) \\ &\qquad\qquad \longrightarrow L(t,x,u,D)u + g(t,x,u) \text{ in } C(I\times M), \end{aligned} \tag{5.1.20}$$

while clearly $\partial u_{\varepsilon_\nu}/\partial t \to \partial u/\partial t$ weakly. Thus (5.1.1) follows in the limit from (5.1.4).

There are questions of uniqueness, stability, and rate of convergence of u_ε to u, which we can treat simultaneously. Thus, with $\varepsilon \in [0,1]$, we compare a solution u to (5.1.1) to a solution u_ε to

$$\frac{\partial u_\varepsilon}{\partial t} = J_\varepsilon L(t,x,J_\varepsilon u_\varepsilon, D)J_\varepsilon u_\varepsilon + J_\varepsilon g(t,x,J_\varepsilon u_\varepsilon), \quad u_\varepsilon(0) = h. \tag{5.1.21}$$

Set

$$v = u - u_\varepsilon$$

and subtract (5.1.21) from (5.1.1). Suppressing the variables (t,x), we have (5.1.22)

$$\partial v/\partial t = L(u,D)v + L(u,D)u_\varepsilon - J_\varepsilon L(J_\varepsilon u_\varepsilon, D)J_\varepsilon u_\varepsilon + g(u) - J_\varepsilon g(J_\varepsilon u_\varepsilon).$$

Write

$$\begin{aligned} L(u,D)u_\varepsilon &- J_\varepsilon L(J_\varepsilon u_\varepsilon, D)J_\varepsilon u_\varepsilon \\ &= [L(u,D) - L(u_\varepsilon, D)]u_\varepsilon \\ &\quad + (1-J_\varepsilon)L(u_\varepsilon,D)u_\varepsilon + J_\varepsilon L(u_\varepsilon,D)(1-J_\varepsilon)u_\varepsilon \\ &\quad + J_\varepsilon[L(u_\varepsilon,D) - L(J_\varepsilon u_\varepsilon, D)]J_\varepsilon u_\varepsilon \end{aligned} \tag{5.1.23}$$

and

$$\begin{aligned} g(u) - J_\varepsilon g(J_\varepsilon u_\varepsilon) = [g(u) - g(u_\varepsilon)] &+ (1-J_\varepsilon)g(u_\varepsilon) \\ &+ J_\varepsilon[g(u_\varepsilon) - g(J_\varepsilon u_\varepsilon)]. \end{aligned} \tag{5.1.24}$$

Now write

$$g(u)-g(w) = G(u,w)(u-w), \quad G(u,w) = \int_0^1 g'(\tau u + (1-\tau)w)\,d\tau, \tag{5.1.25}$$

and similarly

$$L(u, D) - L(w, D) = (u - w) \cdot M(u, w, D). \tag{5.1.26}$$

Then (5.1.22) yields

$$\partial v/\partial t = L(u, D)v + A(u, u_\varepsilon, \nabla u_\varepsilon)v + R_\varepsilon \tag{5.1.27}$$

where

$$A(u, u_\varepsilon, \nabla u_\varepsilon)v = v \cdot M(u, u_\varepsilon, D)u_\varepsilon + G(u, u_\varepsilon)v \tag{5.1.28}$$

incorporates the first terms on the right sides of (5.1.23) and (5.1.24), and R_ε is the sum of the rest of the terms in (5.1.23) - (5.1.24). Note that each term making up R_ε has a factor $I - J_\varepsilon$, acting on either u_ε, $g(u_\varepsilon)$, or $L(u_\varepsilon, D)u_\varepsilon$. Thus there is an estimate

$$\|R_\varepsilon(t)\|_{L^2}^2 \le C_s(\|u_\varepsilon(t)\|_{C^1})(1 + \|u_\varepsilon(t)\|_{H^s}^2)r_s(\varepsilon)^2 \tag{5.1.29}$$

where

$$r_s(\varepsilon) = \|I - J_\varepsilon\|_{\mathcal{L}(H^{s-1}, L^2)} \approx \|I - J_\varepsilon\|_{\mathcal{L}(H^s, H^1)}. \tag{5.1.30}$$

Now, estimating $(d/dt)\|v(t)\|_{L^2}^2$ via the obvious analogue of (5.1.9) yields

$$(d/dt)\|v(t)\|_{L^2}^2 \le C(t)\|v(t)\|_{L^2}^2 + S(t) \tag{5.1.31}$$

with

$$C(t) = C(\|u_\varepsilon(t)\|_{C^1}, \|u(t)\|_{C^1}), \quad S(t) = \|R_\varepsilon(t)\|_{L^2}^2. \tag{5.1.32}$$

Consequently, by Gronwall's inequality, with $K(t) = \int_0^1 C(\tau)d\tau$,

$$\|v(t)\|_{L^2}^2 \le e^{K(t)}\Big[\|f - h\|_{L^2}^2 + \int_0^1 S(\tau)e^{-K(\tau)}d\tau\Big]. \tag{5.1.33}$$

This estimate establishes the following

Proposition 5.1.C. *For $s > \frac{n}{2} + 1$, solutions to (5.1.1) satisfying (5.1.17) are unique. They are limits of solutions u_ε to (5.1.4), and, for $t \in I$,*

$$\|u(t) - u_\varepsilon(t)\|_{L^2} \le K_1(t)\|I - J_\varepsilon\|_{\mathcal{L}(H^{s-1}, L^2)}. \tag{5.1.34}$$

Note that if $J_\varepsilon = \varphi(\varepsilon\Lambda)$ with $\varphi \in C_0^\infty(\mathbb{R})$, $\varphi(\lambda) = 1$ for $|\lambda| < 1$, then we have the operator norm estimate

(5.1.35) $$\|I - J_\varepsilon\|_{\mathcal{L}(H^{s-1},L^2)} \le C \cdot \varepsilon^{s-1}.$$

Returning to properties of solutions to (5.1.1), we establish the following small but significant improvement of (5.1.17).

PROPOSITION 5.1.D. *Given* $f \in H^s(M)$, $s > \frac{n}{2} + 1$, *the solution* u *to* (5.1.1) *satisfies*

(5.1.36) $$u \in C(I, H^s(M)).$$

For the proof, note that (5.1.17) implies that $u(t)$ is a continuous function of t with values in $H^s(M)$, given the weak topology. To establish (5.1.36), it suffices to demonstrate that the norm $\|u(t)\|_{H^s}$ is a continuous function of t. We estimate the rate of change of $\|u(t)\|_{H^s}^2$ by a device similar to the analysis of (5.1.7). Unfortunately, it is not useful to look directly at $(d/dt)\|\Lambda^s u(t)\|_{L^2}^2$, since $L\Lambda^s u$ will not be in L^2. To get around this, we throw in a factor of J_ε, and look at

(5.1.37)
$$\frac{d}{dt}\|\Lambda^s J_\varepsilon u(t)\|_{L^2}^2 = 2(\Lambda^s J_\varepsilon L(u,D)u, \Lambda^s J_\varepsilon u) + 2(\Lambda^s J_\varepsilon g(u), \Lambda^s J_\varepsilon u).$$

As above, we have suppressed the dependence on t, x, for notational convenience. The last term on the right is easy to estimate; we write the first term as

(5.1.38) $$2(\Lambda^s L(u,D)u, \Lambda^s J_\varepsilon^2 u) = 2(L\Lambda^s u, \Lambda^s J_\varepsilon^2 u) + 2([\Lambda^s, L]u, \Lambda^s J_\varepsilon^2 u).$$

Here, for fixed t, $L(u,D)\Lambda^s u \in H^{-1}(M)$, whish can be paired with $\Lambda^s J_\varepsilon^2 u \in C^\infty(M)$. Now we use the Kato-Ponce estimate to obtain

(5.1.39) $$\|[\Lambda^s, L]u\|_{L^2} \le C \sum_j \Big[\|A_j(u)\|_{H^s}\|u\|_{C^1} + \|A_j(u)\|_{C^1}\|u\|_{H^s}\Big],$$

parallel to (5.1.12). This gives control over the last term in (5.1.38). We can write the first term on the right side of (5.1.38) as

(5.1.40) $$((L + L^*)\Lambda^s J_\varepsilon u, \Lambda^s J_\varepsilon u) + 2([J_\varepsilon, L]\Lambda^s u, \Lambda^s J_\varepsilon u).$$

The first term is bounded just as in (5.1.9) - (5.1.10). As for the last term, we have

(5.1.41) $$[J_\varepsilon, L]w = \sum_j [A_j(u), J_\varepsilon]\partial_j w.$$

We have the estimate

$$\|[A_j, J_\varepsilon]\partial_j w\|_{L^2} \leq C\|A_j\|_{C^1}\|w\|_{L^2}, \tag{5.1.42}$$

following by duality from the elementary bound

$$\|[A_j, J_\varepsilon]f\|_{H^1} \leq C\|A_j\|_{C^1}\|f\|_{L^2}. \tag{5.1.43}$$

Consequently we have a bound

$$(d/dt)\|J_\varepsilon u(t)\|_{H^s}^2 \leq C(\|u(t)\|_{C^1})\|u(t)\|_{H^s}^2, \tag{5.1.44}$$

the right side being independent of $\varepsilon \in (0,1]$. Using time-reversal gives a bound on the *absolute value* of the left side of (5.1.44). Thus $\|J_\varepsilon u(t)\|_{H^s}^2 = N_\varepsilon(t)$ is Lipschitz continuous in t, uniformly in ε. As $J_\varepsilon u(t) \longrightarrow u(t)$ in H^s-norm for each $t \in I$, it follows that $\|u(t)\|_{H^s}^2 = N_0(t) = \lim_{\varepsilon \to 0} N_\varepsilon(t)$ has this same Lipschitz continuity. The proof is complete.

It is well known that in general symmetric quasilinear hyperbolic equations might have solutions that break down in finite time. We mention two simple illustrative examples. First consider

$$\partial u/\partial t = u^2, \quad u(0,x) = 1. \tag{5.1.45}$$

The solution is $u(t,x) = (1-t)^{-1}$, which blows up as $t \to 1$. Next, consider

$$u_t + uu_x = 0, \quad u(0,x) = f(x). \tag{5.1.46}$$

We see that $u(t,x)$ is constant on straight lines through $(x,0)$, with slope $f(x)^{-1}$, in the (x,t)-plane. However, it is inevitable that such lines intersect. At the point of first intersection, $u_x(t,x)$ blows up. A shock wave is formed.

We now show that, in a general context, breakdown of a classical solution must involve blow-up of either $\sup_x |u(t,x)|$ or $\sup_x |\nabla_x u(t,x)|$.

PROPOSITION 5.1.E. *Suppose* $u \in C([0,T), H^s(M))$, $s > \frac{n}{2} + 1$, *and assume* u *solves* (5.1.1) *for* $t \in (0,T)$. *Assume also that*

$$\|u(t)\|_{C^1(M)} \leq K < \infty, \tag{5.1.47}$$

for $t \in [0,T)$. *Then there exists* $T_1 > T$ *such that* u *extends to a solution to* (5.1.1), *belonging to* $C([0,T_1), H^s(M))$.

PROOF: This follows easily from the estimate (5.1.44), which has the form $dN_\varepsilon/dt \leq C_1(t)N_0(t)$. If we write this in an equivalent integral form:

$$N_\varepsilon(t+\tau) \leq N_\varepsilon(t) + \int_t^{t+\tau} C_1(s)N_0(s)ds,$$

it is clear that we can pass to the limit $\varepsilon \to 0$, obtaining the differential inequality

$$dN_0/dt \le C(\|u(t)\|_{C^1})N_0(t) \tag{5.1.48}$$

for the Lipschitz function $N_0(t)$. Now Gronwall's inequality implies $N_0(t)$ cannot blow up as $t \to T$ unless $\|u(t)\|_{C^1}$ does, so we are done

This result was established in [Mj] for s an integer. Proving such results for noninteger s was one of the principal motivations for Kato and Ponce to establish their commutator estimate in [KP].

We make the following remark on the factor of the form $C(\|u\|_{C^1})$ that appears in the estimates (5.1.14) and (5.1.44). Namely, a check of the ingredients which produced this factor, such as (5.1.12)-(5.1.13), shows that this factor has *linear* dependence on the C^1-norm, though possibly nonlinear dependence on the sup norm. Hence

$$C(\|u\|_{C^1}) = C_0(\|u\|_{L^\infty})\big[\|u\|_{C^1} + 1\big]. \tag{5.1.49}$$

Using this and an argument similar to one in [BKM], we can sharpen up Proposition 5.1.E a bit.

PROPOSITION 5.1.F. *In the setting of Proposition* 5.1.E, *the solution persists as long as* $\|u(t)\|_{C^1_*(M)}$ *is bounded.*

PROOF: We supplement (5.1.48)-(5.1.49) with the following estimate, which follows from (B.2.12) in Appendix B.

$$\|u(t)\|_{C^1} \le C\|u(t)\|_{C^1_*}\big[1 + \log(\|u\|_{H^s}/\|u\|_{C^1_*})\big], \tag{5.1.50}$$

given $s > \frac{1}{2}n + 1$. Hence, if $\|u(t)\|_{C^1_*}$ is bounded, we have

$$dN_0/dt \le K N_0(t)\big[1 + \log^+ N_0(t)\big], \tag{5.1.51}$$

for $N_0(t) = \|u(t)\|^2_{H^s}$. Since

$$\int_c^\infty \frac{d\tau}{\tau\ \log \tau} = \infty, \tag{5.1.52}$$

Gronwall's inequality applied to this estimate yields a bound on $\|u(t)\|_{H^s}$ for all t for which $\|u(t)\|_{C^1_*} \le K$, completing the proof.

§5.2. Symmetrizable hyperbolic systems

The results of the previous section extend to the case

$$A_0(t,x,u)\frac{\partial u}{\partial t}=\sum_{j=1}^{n}A_j(t,x,u)\partial_j u+g(t,x,u),\quad u(0)=f, \tag{5.2.1}$$

where, as in (5.1.3), all A_j are symmetric, and furthermore

$$A_0(t,x,u)\geq cI\geq 0. \tag{5.2.2}$$

We have:

PROPOSITION 5.2.A. *Given $f\in H^s(M)$, $s>\frac{1}{2}n+1$, the existence and uniqueness results of §5.1 continue to hold for* (5.2.1).

We obtain the solution u to (5.2.1) as a limit of solutions u_ε to

$$A_0(t,x,J_\varepsilon u_\varepsilon)\partial u_\varepsilon/\partial t=J_\varepsilon L_\varepsilon J_\varepsilon u_\varepsilon+g_\varepsilon,\quad u_\varepsilon(0)=f, \tag{5.2.3}$$

where L_ε and g_ε are as in (5.1.5) - (5.1.6). We need to parallel the estimates of §5.1, particularly (5.1.7) - (5.1.14). The key is to replace L^2-inner products by

$$(w,A_{0\varepsilon}(t)w),\quad A_{0\varepsilon}(t)=A_0(t,x,J_\varepsilon u_\varepsilon), \tag{5.2.4}$$

which by hypothesis (5.2.2) will define equivalent L^2 norms. We have

$$\begin{aligned}\frac{d}{dt}&(\Lambda^s u_\varepsilon,A_{0\varepsilon}(t)\Lambda^s u_\varepsilon)\\&=2(\Lambda^s(\partial u_\varepsilon/\partial t),A_{0\varepsilon}(t)\Lambda^s u_\varepsilon)+(\Lambda^s u_\varepsilon,A_{0\varepsilon}'(t)\Lambda^s u_\varepsilon).\end{aligned} \tag{5.2.5}$$

The first term on the right side of (5.2.5) can be written

$$2(\Lambda^s A_{0\varepsilon}(\partial u_\varepsilon/\partial t),\Lambda^s u_\varepsilon)+2([\Lambda^s,A_{0\varepsilon}](\partial u_\varepsilon/\partial t),\Lambda^s u_\varepsilon); \tag{5.2.6}$$

in the first of these terms, we replace $A_{0\varepsilon}(\partial u_\varepsilon/\partial t)$ by the right side of (5.2.3), and estimate the resulting expression by the same method as was applied to the right side of (5.1.7). The commutator $[\Lambda^s,A_{0\varepsilon}]$ is amenable to an estimate parallel to (5.1.11); then substitute for $\partial u_\varepsilon/\partial t$, $A_{0\varepsilon}^{-1}$ times the right side of (5.2.3), and the last term in (5.2.6) is easily estimated. It remains to treat the last term in (5.2.5). We have

$$A_{0\varepsilon}'(t)=(d/dt)A_0(t,x,J_\varepsilon u_\varepsilon(t,x)), \tag{5.2.7}$$

hence

$$\|A'_{0\varepsilon}(t)\|_{L^\infty(M)} \le C(\|J_\varepsilon u_\varepsilon\|_{L^\infty}, \|J_\varepsilon u'_\varepsilon(t)\|_{L^\infty}). \tag{5.2.8}$$

Of course, $\|\partial u_\varepsilon/\partial t\|_{L^\infty}$ can be estimated by $\|u_\varepsilon(t)\|_{C^1}$ by (5.2.3). Consequently, we obtain an estimate parallel to (5.1.14), namely

$$(d/dt)(\Lambda^s u_\varepsilon, A_{0\varepsilon}\Lambda^s u_\varepsilon) \le C_s(\|u_\varepsilon(t)\|_{C^1})(1 + \|J_\varepsilon u_\varepsilon(t)\|^2_{H^s}). \tag{5.2.9}$$

From here, the rest of the parallel with §5.1 is clear.

The class of systems (5.2.1), with all $A_j = A_j^*$ and $A_0 \ge cI > 0$, is an extension of the class of symmetric hyperbolic systems. We call a system

$$\partial u/\partial t = \sum_{j=1}^n B_j(t,x,u)\partial_j u + g(t,x,u), \quad u(0) = f \tag{5.2.10}$$

a symmetrizable hyperbolic system provided there exist $A_0(t,x,u)$, positive definite, such that $A_0(t,x,u)B_j(t,x,u) = A_j(t,x,u)$ are all symmetric. Then applying $A_0(t,x,u)$ to (5.2.10) yields an equation of the form (5.2.1) (with different g and f), so the existence and uniqueness results of §5.1 apply. The factor $A_0(t,x,u)$ is called a symmetrizer.

An important example of such a situation is provided by the equations of compressible fluid flow

$$\begin{aligned} &\partial v/\partial t + \nabla_v v + (1/\rho)\nabla p = 0, \\ &\partial \rho/\partial t + \nabla_v \rho + \rho \text{ div } v = 0. \end{aligned} \tag{5.2.11}$$

Here v is the velocity field of a fluid of density $\rho = \rho(t,x)$. We consider the model in which p is assumed to be a function of ρ. In this situation one says the flow is 'isentropic.' A particular example is

$$p(\rho) = A\rho^\gamma, \tag{5.2.12}$$

with $A > 0,\ 1 < \gamma < 2$; for air, $\gamma = 1.4$ is a good approximation.

The system (5.2.11) is not a symmetric hyperbolic system as it stands. Before constructing a symmetrizer, we will transform it. It is standard practice to rewrite (5.2.11) as a system for (p, v); using (5.2.12) one has

$$\begin{aligned} &\partial p/\partial t + \nabla_v p + (\gamma p) \text{ div } v = 0 \\ &\partial v/\partial t + \nabla_v v + (1/\rho(p))\nabla p = 0. \end{aligned} \tag{5.2.13}$$

This is symmetrizable. Multiplying these two equations by $(\gamma p)^{-1}$ and $\rho(p)$, respectively, we can rewrite the system as

$$\begin{aligned} (\gamma p)^{-1}\partial p/\partial t &= -(\gamma p)^{-1}\nabla_v p - \text{ div } v \\ \rho(p)\partial v/\partial t &= -\nabla p - \rho(p)\nabla_v v. \end{aligned} \tag{5.2.14}$$

Now (5.2.14) is a symmetric hyperbolic system of the form (5.2.1), since, by the divergence theorem,

$$(\mathrm{div}\ v, p)_{L^2(M)} = -(v, \nabla p)_{L^2(M)}. \tag{5.2.15}$$

Thus the results of §5.1 apply to the equations §5.2.13 for compressible fluid flow, as long as p and $\rho(p)$ are bounded away from 0.

Various important second order quasilinear hyperbolic equations can be converted to symmetrizable first order systems. We indicate how this can be done for equations of the form

(5.2.16)
$$\partial_t^2 u - \sum_j B^j(t,x,D^1u)\partial_j\partial_t u - \sum_{j,k} A^{jk}(t,x,D^1u)\partial_j\partial_k u = C(t,x,D^1u).$$

For simplicity we suppose B^j and A^{jk} are scalar. We will produce a first order system for $W = (u, u_0, u_1, \dots, u_n)$, where

$$u_0 = \partial u/\partial t, \quad u_j = \partial_j u, \quad 1 \le j \le n. \tag{5.2.17}$$

We get

$$\begin{aligned} \partial_t u &= u_0 \\ \partial_t u_0 &= \sum B^j(t,x,W)\partial_j u_0 + \sum A^{jk}(t,x,W)\partial_j u_k + C(t,x,W) \\ \partial_t u_j &= \partial_j u_0, \end{aligned} \tag{5.2.18}$$

which is a system of the form

$$\partial W/\partial t = \sum_j H_j(t,x,W)\partial_j W + g(t,x,W). \tag{5.2.19}$$

We can apply to each side of (5.2.19) a matrix of the following form: a block diagonal matrix, consisting of the 2×2 identity matrix in the upper left and the matrix A^{-1} in the lower right, where $A = (A^{jk})$, provided we make the hypothesis that A is positive definite, i.e.,

$$\sum A^{jk}(t,x,W)\xi_j\xi_k \ge C|\xi|^2. \tag{5.2.20}$$

Under this hypothesis, (5.2.19) is symmetrizable. Consequently, we have

PROPOSITION 5.2.B. *Under the hypothesis (5.2.20), if $u(0) = f \in H^{s+1}(M)$, $u_t(0) \in H^s(M), s > \frac{1}{2}n + 1$, then there is a unique local solution*

$$u \in C(I, H^{s+1}(M)) \cap C^1(I, H^s(M)) \tag{5.2.21}$$

to (5.2.16), *which persists as long as*

$$\|u(t)\|_{C^2_\bullet(M)} + \|u_t(t)\|_{C^1_\bullet(M)} \tag{5.2.22}$$

is bounded.

We note that (5.2.20) is stronger than the natural hypothesis of strict hyperbolicity, which is that, for $\xi \neq 0$, the characteristic polynomial

$$\tau^2 - \sum_j B^j(t,x,W)\xi_j\tau - \sum_{j,k} A^{jk}(t,x,W)\xi_j\xi_k = 0 \tag{5.2.23}$$

has two distinct real roots $\tau = \lambda_\nu(t,W,x,\xi)$. However, in the more general strictly hyperbolic case, using Cauchy data to define a Lorentz metric over the initial surface $\{t=0\}$, we can effect a local coordinate change so that, at $t = 0, (A^{jk})$ is positive definite, when the PDE is written in these new coordinates, and then the local existence in Proposition 5.2.B applies.

We now introduce a more general notion of symmetrizer, following Lax [L1], which will bring in pseudodifferential operators. We will say that a function $R(t,u,x,\xi)$, smooth on $\mathbb{R}\times\mathbb{R}^K\times T^*M\setminus 0$, homogeneous of degree 0 in ξ, is a symmetrizer for (5.2.10) provided

$$R(t,u,x,\xi) \text{ is a positive definite } K\times K \text{ matrix} \tag{5.2.26}$$

and

$$R(t,u,x,\xi)\sum B_j(t,x,u)\xi_j \text{ is self adjoint,} \tag{5.2.27}$$

for each (t,u,x,ξ). We then say (5.2.10) is symmetrizable. One reason for the importance of this notion is the following.

PROPOSITION 5.2.C. *Whenever* (5.2.10) *is strictly hyperbolic, it is symmetrizable.*

PROOF: If we denote the eigenvalues of $L(t,u,x,\xi) = \sum B_j(t,x,u)\xi_j$ by $\lambda_1(t,u,x,\xi) < \cdots < \lambda_K(t,u,x,\xi)$, then λ_ν are well defined C^∞ functions of (t,u,x,ξ), homogeneous of degree 1 in ξ. If $P_\nu(t,u,x,\xi)$ are the projections onto the λ_ν-eigenspaces of L^*,

$$P_\nu = \frac{1}{2\pi i}\int_{\gamma_\nu} (\zeta - L(t,u,x,\xi)^*)^{-1}\,d\zeta, \tag{5.2.28}$$

then P_ν is smooth and homogeneous of degree 0 in ξ. Then

$$R(t,u,x,\xi) = \sum_j P_j(t,u,x,\xi)P_j(t,u,x,\xi)^* \tag{5.2.29}$$

gives the desired symmetrizer.

Note that

$$u \in C^{1+r} \Longrightarrow R \in C^{1+r} S^0_{cl}. \tag{5.2.30}$$

Now, with $R = R(t, u, x, D)$, set

$$Q = \frac{1}{2}(R + R^*) + K\Lambda^{-1}, \tag{5.2.31}$$

where $K > 0$ is chosen so that Q is a positive definite operator on L^2.

We will work with approximate solutions u_ε to (5.2.10), given by (5.1.4), with

$$L_\varepsilon v = \sum_j B_j(t, x, J_\varepsilon u_\varepsilon)\partial_j v. \tag{5.2.32}$$

We want to obtain estimates on $(\Lambda^s u_\varepsilon(t), Q_\varepsilon \Lambda^s u_\varepsilon(t))$, where Q_ε arises by the process above, from $R_\varepsilon = R(t, J_\varepsilon u_\varepsilon, x, \xi)$. We begin with

$$\frac{d}{dt}(\Lambda^s u_\varepsilon, Q_\varepsilon \Lambda^s u_\varepsilon) = 2(\Lambda^s(\partial u_\varepsilon/\partial t), Q_\varepsilon \Lambda^s u_\varepsilon) + (\Lambda^s u_\varepsilon, Q'_\varepsilon \Lambda^s u_\varepsilon). \tag{5.2.33}$$

In the last term we can replace Q'_ε by $(d/dt)R(t, J_\varepsilon u_\varepsilon, x, D)$, and obtain

$$|(\Lambda^s u_\varepsilon, Q'_\varepsilon \Lambda^s u_\varepsilon)| \leq C(\|u_\varepsilon(t)\|_{C^1})\|u_\varepsilon(t)\|^2_{H^s}. \tag{5.2.34}$$

We can write the first term on the right side of (5.2.33) as twice

$$(Q_\varepsilon \Lambda^s J_\varepsilon L_\varepsilon J_\varepsilon u_\varepsilon, \Lambda^s u_\varepsilon) + (Q_\varepsilon \Lambda^s g_\varepsilon, \Lambda^s u_\varepsilon.) \tag{5.2.35}$$

The last term has an easy estimate. We write the first term as

$$\begin{aligned}(Q_\varepsilon L_\varepsilon \Lambda^s J_\varepsilon u_\varepsilon, \Lambda^s J_\varepsilon u_\varepsilon) + (Q_\varepsilon[\Lambda^s, L_\varepsilon] J_\varepsilon u_\varepsilon, \Lambda^s J_\varepsilon u_\varepsilon) \\ + ([Q_\varepsilon \Lambda^s, J_\varepsilon] L_\varepsilon J_\varepsilon u_\varepsilon, \Lambda^s u_\varepsilon).\end{aligned} \tag{5.2.36}$$

Note that, as long as (5.2.30) holds, with $r \geq 0$, R_ε also has symbol in $C^{1+r} S^0_{cl}$, and we have, by Proposition 4.1.E,

$$[Q_\varepsilon \Lambda^s, J_\varepsilon] \text{ bounded in } \mathcal{L}(H^{s-1}, L^2), \tag{5.2.37}$$

with bound given in terms of $\|u_\varepsilon(t)\|_{C^1}$. Now Moser estimates yield

$$\|L_\varepsilon J_\varepsilon u_\varepsilon\|_{H^{s-1}} \leq C(\|u_\varepsilon\|_{L^\infty})\|u_\varepsilon\|_{H^s} + C(\|u_\varepsilon\|_{C^1})\|u_\varepsilon\|_{H^{s-1}}. \tag{5.2.38}$$

Consequently we deduce

$$(5.2.39) \qquad |([Q_\varepsilon \Lambda^s, J_\varepsilon] L_\varepsilon J_\varepsilon u_\varepsilon, \Lambda^s u_\varepsilon)| \leq C(\|u_\varepsilon(t)\|_{C^1}) \|u_\varepsilon(t)\|_{H^s}^2.$$

Moving to the second term in (5.2.36), note that, for $L = \sum B_j(t, x, u)\partial_j$,

$$(5.2.40) \qquad [\Lambda^s, L] = \sum_j [\Lambda^s, B_j(t, x, u)] \partial_j v.$$

By the Kato-Ponce estimate, as in (5.1.12), we have

$$(5.2.41) \qquad \big\| [\Lambda^s, L] v \big\|_{L^2} \leq C \sum_j \Big[\|B_j\|_{Lip^1} \|v\|_{H^s} + \|B_j\|_{H^s} \|v\|_{Lip^1} \Big].$$

Hence the second term in (5.2.36) is also bounded by $C(\|u_\varepsilon\|_{C^1}) \|u_\varepsilon\|_{H^s}^2$.

It remains to estimate the first term in (5.2.36). We claim that

$$(5.2.42) \qquad (QLv, v) \leq C(\|u\|_{C^1}) \|v\|_{L^2}^2.$$

We will obtain this from results of Chapter 4. By Proposition 4.2.C,

$$(5.2.43) \qquad R - R^* : H^s \longrightarrow H^{s+1}, \quad -1 \leq s \leq 0,$$

so in (5.2.42) we can replace Q by R. By Proposition 4.2.A,

$$(5.2.44) \qquad RL = C(x, D) + S$$

where $C(x, \xi) \in C^1 S^1_{cl}$ is i times the symbol (5.2.27), hence is skew-adjoint, and $S : L^2 \longrightarrow L^2$. Finally, the estimate

$$(5.2.45) \qquad (C(x, D)v, v) \leq C\|v\|_{L^2}^2$$

follows from Proposition 4.2.B.

Our analysis of (5.2.33) is complete; we have

$$(5.2.46) \qquad (d/dt)(\Lambda^s u_\varepsilon, Q_\varepsilon \Lambda^s u_\varepsilon) \leq C(\|u_\varepsilon(t)\|_{C^1}) \|u_\varepsilon(t)\|_{H^s}^2.$$

From here we can parallel the rest of the argument of §5.1, to prove the following.

THEOREM 5.2.D. *If (5.2.10) is symmetrizable, in particular if it is strictly hyperbolic, the initial value problem, with $u(0) = f \in H^s(M)$, has a unique local solution $u \in C(I, H^s(M))$, whenever $s > \frac{1}{2}n + 1$, which persists as long as $\|u(t)\|_{C^1_*}$ is bounded.*

§5.3. Higher order hyperbolic equations

In §5.2 we reduced a fairly general class of second order hyperbolic equations to first order symmetrizable systems. Here we treat equations of degree m, making heavier use of pseudodifferential operators. Consider a quasilinear equation

$$\partial_t^m u = \sum_{j=0}^{m-1} A_j(t,x,D^{m-1}u,D_x)\partial_t^j u + C(t,x,D^{m-1}u), \tag{5.3.1}$$

with initial conditions

$$u(0) = f_0, \partial_t u(0) = f_1, \dots, \partial_t^{m-1}u(0) = f_{m-1}. \tag{5.3.2}$$

Here, $A_j(t,x,w,D_x)$ is a differential operator, homogeneous of degree $m-j$. Assume u takes values in $\mathbb{R}^K$, but for simplicity we suppose A_j have scalar coefficients. We will produce a first order system for $v = (v_0, \dots, v_{m-1})$ with

$$v_0 = \Lambda^{m-1}u, \dots, v_j = \Lambda^{m-j-1}\partial_t^j u, \dots, v_{m-1} = \partial_t^{m-1}u. \tag{5.3.3}$$

We have

$$\begin{aligned} \partial_t v_0 &= \Lambda v_1 \\ &\vdots \\ \partial_t v_{m-2} &= \Lambda v_{m-1} \\ \partial_t v_{m-1} &= \sum A_j(t,x,Pv,D_x)\Lambda^{1+j-m}v_j + C(t,x,Pv), \end{aligned} \tag{5.3.4}$$

where $Pv = D^{m-1}u$, i.e., $\partial_x^\beta \partial_t^j u = \partial_x^\beta \Lambda^{j+1-m} v_j$, so $P \in OPS^0_{cl}$. Note that $A_j(t,x,Pv,D_x)\Lambda^{1+j-m}$ is an operator of order 1. The initial condition is

$$v_0(0) = \Lambda^{m-1}f_0, \dots, v_j(0) = \Lambda^{m-j-1}f_j, \dots, v_{m-1}(0) = f_{m-1}. \tag{5.3.5}$$

The system (5.3.4) has the form

$$\partial_t v = L(t,x,Pv,D)v + G(t,x,Pv), \tag{5.3.6}$$

where L is an $m \times m$ matrix of pseudodifferential operators, which are scalar (though each entry acts on K-vectors). Quasilinear hyperbolic pseudodifferential equations like this were studied in §4.5 of [T2], though with a less precise analysis than we give here. Note that the eigenvalues of the principal symbol of L are $i\lambda_\nu(t,x,v,\xi)$, where $\tau = \lambda_\nu$ are the roots of the characteristic equation

$$\tau^m - \sum_{j=0}^{m-1} A_j(t,x,Pv,\xi)\tau^j = 0. \tag{5.3.7}$$

We will make the hypothesis of strict hyperbolicity, that for $\xi \neq 0$ this equation has m distinct real roots, so $L(t,x,Pv,\xi)$ has m distinct purely imaginary eigenvalues. Consequently, as in Proposition 5.2.B, there exists a symmetrizer, an $m \times m$ matrix valued function $R(t,x,w,\xi)$, homogeneous of degree 0 in ξ and smooth in its arguments, such that, for $\xi \neq 0$,

$$\begin{aligned} &R(t,x,w,\xi) \text{ is positive definite,} \\ &R(t,x,w,\xi)L(t,x,w,\xi) \text{ is skew adjoint.} \end{aligned} \tag{5.3.8}$$

Note that

$$\begin{aligned} Pv \in C^1 \Longrightarrow L(t,x,Pv,\xi) &\in C^1 S^1_{cl} \text{ and} \\ R(t,x,Pv,\xi) &\in C^1 S^0_{cl}. \end{aligned} \tag{5.3.9}$$

Since the operators $\partial_j \Lambda^{-1}$ do not preserve C^1, we need to work with a slightly smaller space, $C^1_{\#}$, as in (4.1.30) with the property that

$$P \in OPS^0_{cl} \Longrightarrow P : C^1_{\#} \longrightarrow C^1. \tag{5.3.10}$$

Now the analysis of symmetrizable systems in §5.2, involving (5.2.32)-(5.2.46), can be extended to this case, with one extra complication, namely the form of $[\Lambda^s, L]$ is now more complicated than (5.2.40). Instead of the Kato-Ponce estimate, we make use of Proposition 4.1.F. If we write

$$L(t,x,Pv,D)v = \sum_{j=1}^{n} B_j(t,x,D)\partial_j v + B_0(t,x,D)v \tag{5.3.11}$$

with $B_\ell \in OPC^1 S^0_{cl}$, then we replace the estimate (5.2.41) by

$$\|[\Lambda^s, L]v\|_{L^2} \leq C \sum_j \Big[\pi^0_{N,C^1}(B_j)\|v\|_{H^s} + \pi^0_{N,H^s}(B_j)\|v\|_{C^1_{\#}}\Big]. \tag{5.3.12}$$

Thus we derive the following analogue of the estimate (5.2.46):

$$(d/dt)(\Lambda^s v_\varepsilon, Q_\varepsilon \Lambda^s v_\varepsilon) \leq C(\|v_\varepsilon(t)\|_{C^1_{\#}})\|v_\varepsilon(t)\|^2_{H^s}. \tag{5.3.13}$$

We have the following result.

THEOREM 5.3.A. *If (5.3.1) is strictly hyperbolic, with initial data* $f_j \in H^{s+m-1-j}(M), s > \frac{1}{2}n + 1$, *then there is a unique local solution*

$$u \in C(I, H^{s+m-1}(M)) \cap C^{m-1}(I, H^s(M)),$$

which persists as long as $\|u(t)\|_{C_*^m} + \|u_t(t)\|_{C_*^{m-1}} + \cdots + \|\partial_t^{m-1} u(t)\|_{C_*^1}$ *is bounded.*

There is one point it remains to establish to have a proof of this result. Namely we need to justify the use of the C_*^{m-j}-norm of $\partial_t^j u(t)$ rather than the stronger $C_\#^{m-j}$-norms in the statement of a sufficient condition for persistence of the solution. An argument similar to that used in the proof of Proposition 5.1.F will yield this.

Parallel to (5.1.48) we have the estimate

$$\frac{d}{dt}\|v(t)\|_{H^s}^2 \leq C(\|v(t)\|_{C_\#^1})\|v(t)\|_{H^s}^2. \tag{5.3.14}$$

Furthermore, parallel to (5.1.49), we have

$$C(\|v(t)\|_{C_\#^1}) = C_0(\|v(t)\|_{C_\#^0})\big[\|v(t)\|_{C_\#^1} + 1\big]. \tag{5.3.15}$$

Now the $C_\#^0$-norm is weaker than the C^r-norm for any $r > 0$, so the first factor on the right side of (5.3.15) is harmless. Meanwhile, the boundedness of $\|\partial_t^j u(t)\|_{C_*^{m-j}}$ is equivalent to the boundedness of

$$\|v(t)\|_{C_*^1} = Q(t) \tag{5.3.16}$$

Now the estimate (B.2.12) in Appendix B implies that, if $s > \frac{1}{2}n + 1$,

$$\|v(t)\|_{C_\#^1} \leq CQ(t)\big[1 + \log^+ \|v(t)\|_{H^s}^2\big]. \tag{5.3.17}$$

Consequently, as long as $Q(t) \leq K$, (5.3.14) yields the differential inequality

$$\frac{d}{dt}\|v(t)\|_{H^s}^2 \leq K_1\|v(t)\|_{H^s}^2\big[1 + \log^+ \|v(t)\|_{H^s}^2\big]. \tag{5.3.18}$$

Now, by (5.1.52), Gronwall's inequality applied to this estimate yields a bound on $\|v(t)\|_{H^s}$ for all $t \geq 0$ for which $Q(t) \leq K$. This completes the proof of Theorem 5.3.A.

Sometimes equations of the form (5.3.1) arise in which the coefficients A_j depend on u but not on all derivatives of order $\leq m - 1$. We consider second order equations of this nature, i.e., of the form

$$\partial_t^2 u = \sum A^{jk}(t, x, u)\partial_j\partial_k u + \sum B^j(t, x, u)\partial_j\partial_t u + C(t, x, D^1 u). \tag{5.3.19}$$

Then the reduction (5.3.3)-(5.3.4) leads to a special case of (5.3.6):

$$\partial_t v = L(t, x, P_1 v, D)v + G(t, x, Pv) \tag{5.3.20}$$

with

$$P \in OPS^0_{cl}, \quad P_1 \in OPS^{-1}_{cl}. \tag{5.3.21}$$

Consequently, the symmetrizer (5.3.8) belongs to $C^1 S^1_{cl}$ as long as $P_1 v \in C^1$, hence as long as $v \in C^0_{\#}$. Thus we can hope to improve the factor $C(\|v_\varepsilon\|_{C^1_{\#}})$ in the estimate (5.3.13). At first glance, the last term in the estimate (5.3.12) presents a problem. However, if we write

$$[\Lambda^s, B_j]\partial_j v = [\Lambda^s \partial_j, B_j]v - \Lambda^s[\partial_j, B_j]v, \tag{5.3.22}$$

we overcome this. Therefore, for the system (5.3.15) we obtain the improvement on (5.3.13):

$$(d/dt)(\Lambda^s v_\varepsilon, Q_\varepsilon \Lambda^s v_\varepsilon) \le C(\|v_\varepsilon(t)\|_{C^0_{\#}})\|v_\varepsilon(t)\|^2_{H^s}, \tag{5.3.23}$$

leading to the following improvement of Theorem 5.3.A.

PROPOSITION 5.3.B. *If (5.3.19) is strictly hyperbolic, with initial data* $f \in H^{s+1}(M), g \in H^s(M), s > \frac{1}{2}n$, *then there is a unique local solution* $u \in C(I, H^{s+1}(M)) \cap C^1(I, H^s(M))$, *which persists as long as* $\|u(t)\|_{C^1_*} + \|u_t(t)\|_{C^0_*}$ *is bounded.*

This result is also established in [HKM], with the exception of the final statement on persistence of the solution. Note that Proposition 5.2.B applies to equations of the form (5.3.19), but it yields a result which is cruder than that of Proposition 5.3.B.

One way in which systems of the form (5.3.19) arise is in the Einstein equations, relating the Ricci tensor on a Lorentz manifold and the stress-energy tensor. As in the Riemannian case (2.2.48), if (h_{jk}) is the Lorentz metric tensor, and if one uses local harmonic coordinates, the Ricci tensor is given by

$$-\frac{1}{2}\sum_{j,k} h^{jk}\partial_j\partial_k h_{\ell m} + Q_{\ell m}(h, Dh) = R_{\ell m}. \tag{5.3.24}$$

Thus one obtains a Cauchy problem for the components of the metric tensor of the form (5.3.19), and Proposition 5.3.B can be applied. This approach to the Einstein equations is classical, going back to C.Lanczos in 1922 and used by a number of mathematicians; see [HKM]. On the other hand, there are reasons to have the flexibility to use other coordinate systems, discussed in [DeT] and in recent work of [CK].

In [FM], the Einstein equations were reduced to a first order system by such a device as we noted in §5.2. The improvement coming from a different treatment of the second order system was noted in [HKM].

§5.4. Completely nonlinear hyperbolic systems

Consider the Cauchy problem for a completely nonlinear first order system

$$\partial u/\partial t = F(t, x, D_x^1 u), \quad u(0) = f. \tag{5.4.1}$$

We assume u takes values in $\mathbb{R}^K$. We then form a first order system for $v = (v_0, v_1, \dots, v_n) = (u, \partial_1 u, \dots, \partial_n u)$.

$$\begin{aligned} \partial v_0/\partial t &= F(t, x, v) \\ \partial v_j/\partial t &= \sum_\ell (\partial_{v_\ell} F)(t, x, v)\partial_\ell v_j + (\partial_{x_j} F)(t, x, v) \ (j \geq 1), \end{aligned} \tag{5.4.2}$$

with initial data

$$v(0) = (f, \partial_1 f, \dots, \partial_n f). \tag{5.4.3}$$

The behavior of this system is controlled by the operator with $K \times K$ matrix coefficients

$$\begin{aligned} L(t, x, v, D) &= \sum_\ell (\partial_{v_\ell} F)(t, x, v)\partial_\ell \\ &= \sum_\ell B_\ell(t, x, v)\partial_\ell. \end{aligned} \tag{5.4.4}$$

We see that (5.4.2) is symmetric hyperbolic if

$$L(t, x, v, \xi) = \sum_\ell B_\ell(t, x, v)\xi_\ell \tag{5.4.5}$$

is a symmetric $K \times K$ matrix, and symmetrizable if there is a symmetrizer, of the form (5.2.26)-(5.2.27), for L. In these respective cases we say (5.4.1) is a symmetric (or symmetrizable) hyperbolic system. We also say (5.4.1) is strictly hyperbolic if, for each $\xi \neq 0$, (5.4.5) has K distinct real eigenvalues; such equations are symmetrizable.

The following is a simple consequence of Theorem 5.2.D.

PROPOSITION 5.4.A. *If (5.4.1) is a symmetrizable hyperbolic system and $f \in H^s(M)$ with $s > \frac{1}{2}n + 2$, then there is a unique local solution $u \in$*

$C(I, H^s(M))$. This solution persists as long as $\|u(t)\|_{C^2} + \|\partial_t u(t)\|_{C^1}$ is bounded.

Similarly consider the Cauchy problem for a completely nonlinear second order equation

$$u_{tt} = F(t, x, D^1 u, \partial_x^1 u_t, \partial_x^2 u), \quad u(0) = f, \ u_t(0) = g. \tag{5.4.6}$$

Here $F = F(t, x, \xi, \eta, \zeta)$ is smooth in its arguments; $\zeta = (\zeta_{jk}) = (\partial_j \partial_k u)$, etc. As before, set $v = (v_0, v_1, \dots, v_n) = (u, \partial_1 u, \dots, \partial_n u)$. We obtain for v a quasilinear system of the form

$$\begin{aligned} \partial_t^2 v_0 &= F(t, x, D^1 v) \\ \partial_t^2 v_i &= \sum_{j,k} (\partial_{\zeta_{jk}} F)(t, x, D^1 v) \partial_j \partial_k v_i \\ &\qquad + \sum_j (\partial_{\eta_j} F)(t, x, D^1 v) \partial_j \partial_t v_i + G_i(t, x, D^1 v), \end{aligned} \tag{5.4.7}$$

with initial data

$$v(0) = (f, \partial_1 f, \dots, \partial_n f), \ v_t(0) = (g, \partial_1 g, \dots, \partial_n g). \tag{5.4.8}$$

The system (5.4.7) is not quite of the form (5.2.16) studied in §5.2, but the difference is minor. One can construct a symmetrizer in the same fashion, as long as

$$\tau^2 = \sum (\partial_{\zeta_{jk}} F)(t, x, D^1 v) \xi_j \xi_k + \sum (\partial_{\eta_j} F)(t, x, D^1 v) \xi_j \tau \tag{5.4.9}$$

has two distinct real roots τ for each $\xi \neq 0$. This is the strict hyperbolicity condition. Proposition 5.2.B holds also for (5.4.7), so we have:

PROPOSITION 5.4.B. *If (5.4.6) is strictly hyperbolic, then given*

$$f \in H^{s+1}(M), \ g \in H^s(M), \ s > \frac{1}{2}n + 2,$$

there is locally a unique solution

$$u \in C(I, H^{s+1}(M)) \cap C^1(I, H^s(M)).$$

This solution persists as long as $\|u(t)\|_{C^3} + \|u_t(t)\|_{C^2}$ is bounded.

This proposition applies to the equations of prescribed Gaussian curvature, for a surface S which is the graph of $y = u(x), x \in \Omega \subset \mathbb{R}^n$, under certain circumstances. The Gauss curvature $K(x)$ is related to $u(x)$ via the PDE

$$\det H(u) - K(x)\big(1 + |\nabla u|^2\big)^{(n+2)/2} = 0, \tag{5.4.10}$$

where $H(u)$ is the Hessian matrix,

$$H(u) = (\partial_j \partial_k u). \tag{5.4.11}$$

Note that, if $F(u) = \det H(u)$, then

$$DF(u)v = Tr[\mathcal{C}(u)H(v)] \tag{5.4.12}$$

where $\mathcal{C}(u)$ is the cofactor matrix of $H(u)$, so

$$H(u)\mathcal{C}(u) = [\det H(u)]I. \tag{5.4.13}$$

Of course, (5.4.10) is elliptic if $K > 0$. Suppose K is negative and on the hypersurface $\Sigma = \{x_n = 0\}$ Cauchy data are prescribed, $u = f(x')$, $\partial_n u = g(x')$, $x' = (x_1, \ldots, x_{n-1})$. Then $\partial_k \partial_j u = \partial_k \partial_j f$ on Σ for $1 \le j, k \le n-1$, $\partial_n \partial_j u = \partial_j g$ on Σ for $1 \le j \le n$, and then (5.4.10) uniquely specifies $\partial_n^2 u$, hence $H(u)$, on Σ, provided $\det H(f) \ne 0$. If the matrix $H(u)$ has signature $(n-1, 1)$, and if Σ is spacelike for its quadratic form, then (5.4.10) is hyperbolic, and Proposition 5.4.B applies.

Chapter 6: Propagation of singularities

We present a proof of Bony's propagation of singularities result for solutions to nonlinear PDE. As mentioned in the Introduction, we emphasize how C^r regularity of solutions rather than $H^{\frac{n}{2}+r}$ regularity yields propagation of higher order microlocal regularity, giving in that sense a slightly more precise result than usual. Our proof also differs from most in using $S^m_{1,\delta}$ calculus, with $\delta < 1$. This simplifies the linear analysis to some degree, but because of this, in another sense our result is slightly weaker than that obtained using $\mathcal{B}^r S^m_{1,1}$ calculus by Bony and Meyer; see also Hörmander's treatment [H4] using $\tilde{S}^m_{1,1}$ calculus. Material developed in §3.4 could be used to supplement the arguments of §6.1, yielding this more precise result. In common with other approaches, our argument is modeled on Hörmander's classic analysis of the linear case.

In §6.2 we give examples of extra singularities, produced by nonlinear interactions rather than by the Hamiltonian flow, and discuss a little the mechanisms behind their creation. It was the discovery of this phenomenon by [La] and [RR] which generated interest in the nonlinear propagation of singularities treated in this chapter.

In §6.3 we discuss a variant of Egorov's theorem for paradifferential operators.

We do not treat reflection of singularities by a boundary, though much interesting work has been done there, by [DW], [Lei], [ST], and others. A description of progress made on such problems can be found in [Be2].

§6.1. Propagation of singularities

Suppose $u \in C^{m+r}(\Omega)$ solves the nonlinear PDE

$$F(x, D^m u) = f. \tag{6.1.1}$$

We assume $r > 0$; this assumption will be strengthened below. We discuss here Bony's result on propagation of singularities [Bo]. Using (3.3.6)-(3.3.10), we have the operator $M(u; x, D) = M^\#(x, D) + M^b(x, D) = M^\# + M^b$, and u satisfies the equation

$$M^\# u = f + R, \tag{6.1.2}$$

where $R = -M^b u \bmod C^\infty$, with $M^b \in OPS^{m-\delta r}_{1,1}$, hence

$$R \in C^{r+\delta r}; \text{ also } u \in H^{m+\sigma,p} \Longrightarrow R \in H^{\sigma+r\delta,p}, \text{ if } \sigma \geq r. \tag{6.1.3}$$

Recall from (3.3.9) that

$$M^\# \in OPA^r_0 S^m_{1,\delta} \subset OPS^m_{1,\delta} \cap OPC^r S^m_{1,0} \tag{6.1.4}$$

and, if $r = \ell + \sigma$, $\ell \in \mathbb{Z}^+$, $0 < \sigma < 1$, then, on the symbol level,

$$D_x^\beta M^\# \in S^m_{1,\delta} \text{ for } |\beta| \le \ell, \quad D_x^\beta M^\# \in S^{m+1-\sigma}_{1,\delta} \text{ for } |\beta| = \ell + 1. \tag{6.1.5}$$

We will use the symbol smoothing decomposition with $\delta < 1$.

To study propagation of singularities for solutions to (6.1.1), we hence study it for solutions to the linear equation (6.1.2). Our analysis will follow Hörmander's well known argument, with some modifications due to the fact that $M^\#$ is not a pseudodifferential operator of classical type. In another context, such a variant arose in [T3]. We will set $\Lambda = (1 - \Delta)^{\frac{1}{2}}$ and

$$P = i^m \; M^\# \Lambda^{1-m} \in OP\mathcal{A}_0^r S^1_{1,\delta} \subset OPS^1_{1,\delta} \cap OPC^r S^1_{1,0}. \tag{6.1.6}$$

More generally, for $\sigma \in \mathbb{R}$, set

$$P_\sigma = \Lambda^\sigma P \Lambda^{-\sigma} \in OPS^1_{1,\delta}. \tag{6.1.7}$$

Using the properties (6.1.5), we have the following.

LEMMA 6.1.A. *Suppose $r > 1$. Set*

$$P = A + iB, \; A = A^*, \; B = B^*. \tag{6.1.8}$$

Then $B \in OPS^0_{1,\delta}$. Furthermore, for each $\sigma \in \mathbb{R}$,

$$P - P_\sigma \in OPS^0_{1,\delta}. \tag{6.1.9}$$

If $a(x,\xi)$ is the real part of the complete symbol of A, given in Lemma 6.1.A, then (provided $r > 1$),

$$a(x,\xi) \in S^1_{1,\delta}, \quad A - a(x,D) \in OPS^0_{1,\delta}, \tag{6.1.10}$$

and, if $r = \ell + \sigma$, $0 < \sigma < 1$,

$$D_x^\beta a(x,\xi) \in S^1_{1,\delta} \text{ for } |\beta| \le \ell, \; D_x^\beta a(x,\xi) \in S^{2-\sigma}_{1,\delta} \text{ for } |\beta| = \ell + 1. \tag{6.1.11}$$

We will examine propagation of singularities for P, but we will want to interpret the results in terms of the symbol of the linearization of the operator $F(x, D^m u)$. Thus, set

$$\tilde{a}(x,\xi) = \sum_{|\alpha| \le m} (\partial F / \partial \zeta_\alpha)(x, D^m u) \xi^\alpha \langle \xi \rangle^{1-m}. \tag{6.1.12}$$

We need to compare $\tilde{a}(x,\xi)$ and $a(x,\xi)$. Note that

$$\tilde{a}(x,\xi) \in C^r S^1_{1,0}, \tag{6.1.13}$$

and if we use the smoothing method of §1.3 to write

$$\tilde{a}(x,\xi) = \tilde{a}^\#(x,\xi) + \tilde{a}^b(x,\xi) \tag{6.1.14}$$

with

$$\tilde{a}^\#(x,\xi) \in S^1_{1,\delta}, \quad \tilde{a}^b(x,\xi) \in C^r S^{1-\delta r}_{1,\delta}, \tag{6.1.15}$$

then, as a consequence of (3.3.14) and Lemma 6.1.A,

$$a(x,\xi) - \tilde{a}^\#(x,\xi) \in S^{1-\rho}_{1,\delta}, \quad \rho = \min(\delta r, 1). \tag{6.1.16}$$

The analysis of propagation of singularities, following Hörmander [H1], begins with the *basic commutator identity*, assuming (6.1.8) holds:

$$\text{Im } (CPu, Cu) = \text{Re } \big(\{(1/i)C^*[C,A] + C^*BC + C^*[B,C]\}u, u\big). \tag{6.1.17}$$

If B is bounded on L^2, then, using

$$|(CPu,Cu)| \le \|CPu\|^2 + \frac{1}{4}\|Cu\|^2,$$

we get the *basic commutator inequality*:

$$\text{Re } \big(\{(1/i)C^*[C,A] - MC^*C\}u, u\big) \le \|CPu\|^2 + |(Wu,u)|, \tag{6.1.18}$$

where

$$M = \|B\| + \frac{1}{4}, \quad W = \text{Re } C^*[B,C]. \tag{6.1.19}$$

C will be a pseudodifferential operator, described more fully below. By convention,
$\text{Re } T = \frac{1}{2}(T + T^*)$.

Now to establish microlocal regularity of u in a conic neighborhood of a certain curve γ in $T^*\Omega \setminus 0$, the strategy is to construct C and $\varphi(x,D)$ of order μ, such that $\varphi(x,\xi)$ is elliptic on γ and, roughly,

$$\text{Re } \{(1/i)C^*[C,A] - MC^*C\}(x,\xi) - \varphi(x,\xi)^2 + E(x,\xi)^2 \ge 0, \tag{6.1.20}$$

for a certain $E(x,\xi) \in S^\mu$ supported on a conic neighborhood of one endpoint of γ. Then we want to apply the sharp Gårding inequality to (6.1.20), and use the result together with (6.1.18) to estimate $\varphi(x, D)u$ in terms of CPu, $E(x, D)u$, and a small additional term.

Arranging (6.1.20) brings in the Poisson bracket and Hamiltonian vector field, defined by

$$
\begin{aligned}
H_a c &= \{c, a\}(x, \xi) \\
&= \sum\Big(\frac{\partial c}{\partial x_j}\frac{\partial a}{\partial \xi_j} - \frac{\partial c}{\partial \xi_j}\frac{\partial a}{\partial x_j}\Big),
\end{aligned}
\tag{6.1.21}
$$

since $(1/i)[C, A]$ has $\{c, a\}(x, \xi)$ for a principal symbol, in a sense which, for the symbols considered here, will be made precise in Lemma 6.1.B below. The curve γ alluded to above will be an integral curve of $H_{\tilde{a}}$.

Following Hörmander, we produce the symbol $c(x, \xi)$ of C in the form

$$
c(x,\xi) = c_{\lambda,\varepsilon}(x,\xi) = d(x,\xi)e^{\lambda f(x,\xi)}\Big(1 + \varepsilon^2 g(x,\xi)^2\Big)^{-\frac{1}{2}}, \tag{6.1.22}
$$

where $\lambda > 0$ will be taken large (fixed) and ε small (tending to 0). Note that

$$
H_{\tilde{a}} c^2 - Mc^2 = 2d_\varepsilon(H_{\tilde{a}} d_\varepsilon)e^{2\lambda f} + d_\varepsilon^2 e^{2\lambda f}[2\lambda H_{\tilde{a}} f - M] \tag{6.1.23}
$$

where

$$
d_\varepsilon = d\langle \varepsilon g\rangle^{-1}
$$

and

$$
H_{\tilde{a}} d_\varepsilon = \langle \varepsilon g\rangle^{-1}\big[H_{\tilde{a}} d - \varepsilon^2\langle \varepsilon g\rangle^{-2}(H_{\tilde{a}} g)d\big]. \tag{6.1.24}
$$

With these calculations in mind, we impose the following properties on d, f, g :

$$
d(x,\xi) \in S^\mu, \ f(x,\xi) \in S^0, \ g(x,\xi) \in S^1, \tag{6.1.25}
$$

all homogeneous for $|\xi|$ large,

$$
H_{\tilde{a}} f \geq 1, \ H_{\tilde{a}} d \geq 0 \text{ (except near } q), \ H_{\tilde{a}} g \leq 0, \ d \geq 0, \tag{6.1.26}
$$

and

$$
H_{\tilde{a}} \geq |\xi|^\mu \text{ on } \Gamma, \ d \text{ supported in } \tilde{\Gamma}, \ g \text{ elliptic on } \tilde{\Gamma}, \tag{6.1.27}
$$

for $|\xi| \geq 1$, where Γ is a small conic neighborhood of the integral curve γ, running from p to q, and $\tilde{\Gamma}$ a slightly bigger conic neighborhood. Here (by slight abuse of notation) we let $H_{\tilde{a}}$ denote the Hamiltonian vector field associated with the principal part $\tilde{a}_1(x,\xi)$ of the symbol (6.1.12). We assume $r > 2$, so $H_{\tilde{a}}$ is a C^1-vector field, with well-defined integral curves. It is easy to arrange (6.1.25) - (6.1.27), in such a fashion that, if $H_{\tilde{a}}$ flows from p to q in γ, then d vanishes on a conic surface Σ transversal to γ, intersecting γ at a point just before p. Having first made $H_{\tilde{a}}d \geq 0$ everywhere on the integral curves through Σ, make a modification in $d(x,\xi)$, cutting it off on a conic neighborhood of q. Thus one has (for λ large)

$$(c\{c,\tilde{a}\} - Mc^2)(x,\xi) \geq \varphi_\varepsilon(x,\xi)^2 - E(x,\xi)^2 \tag{6.1.28}$$

where $\varphi_\varepsilon(x,\xi) = \varphi(x,\xi)\langle\varepsilon g\rangle^{-1}$, with $\varphi, E \in S^\mu$, φ elliptic on γ, and E supported on a small conic neighborhood of q.

Now, to relate this to (6.1.20), we need to replace $\tilde{a}$ by $a(x,\xi)$, the real part of the complete symbol of A, given by Lemma 6.1.A. Then (provided $r > 1$), we recall that $a(x,\xi)$ satisfies (6.1.10); furthermore, if $r = \ell+\sigma$, $0 < \sigma < 1$,

$$D_x^\beta a(x,\xi) \in S^1_{1,\delta} \text{ for } |\beta| \leq \ell, \quad D_x^\beta a(x,\xi) \in S^{2-\sigma}_{1,\delta} \text{ for } |\beta| = \ell+1. \tag{6.1.29}$$

Using this, we deduce the following.

LEMMA 6.1.B. *If $r > 1$ and $C \in OPS^\mu_{1,0}$, then*

$$\{c,a\}(x,\xi) \in S^\mu_{1,\delta} \cap C^{r-1}S^\mu_{1,0} \tag{6.1.30}$$

and

$$Re\ ((1/i)C^*[C,A])(x,\xi) - c\{c,a\}(x,\xi) \in S^{2\mu-(2-2\delta)}_{1,\delta}. \tag{6.1.31}$$

If $r > 2$, the difference (6.1.31) belongs to $S^{2\mu-(2-\delta)}_{1,\delta}$.

PROOF: Straightforward check of symbol expansions.

By (6.1.14) - (6.1.16), we have $a - \tilde{a} \in C^r S^{1-\rho}_{1,\delta}$, where $\rho = \min(\delta r, 1)$, hence, given $C \in OPS^\mu_{1,0}$,

$$|\{c,a\}(x,\xi) - \{c,\tilde{a}\}(x,\xi)| \leq K\langle\xi\rangle^{\mu-\rho}, \text{ if } r > 1. \tag{6.1.32}$$

We therefore have the following rigorous version of the result stated loosely in (6.1.20):

LEMMA 6.1.C. *Assume $r > 2$, and let γ be an integral curve of $H_{\tilde{a}}$, from p to q, Γ a small conic neighborhood of γ, $\tilde{\Gamma}$ a bigger conic neighborhood.*

There exists $c_\varepsilon(x,\xi)$, *bounded in* $S^\mu_{1,0}$, *of the form* (6.1.21), *such that, with* $C = C_\varepsilon = c_\varepsilon(x,D)$,

$$c\{c,a\}(x,\xi) - Mc(x,\xi)^2 - \varphi_\varepsilon(x,\xi)^2 + E(x,\xi)^2 \geq -K\langle\xi\rangle^{2\mu-\rho}, \tag{6.1.33}$$

where $E(x,\xi) \in S^\mu$ *is supported on a small conic neighborhood of* q, $c(x,\xi)$ *is supported on* $\tilde{\Gamma}$, *and*

$$\varphi_\varepsilon(x,\xi) = \langle\varepsilon g(x,\xi)\rangle^{-1}\varphi(x,\xi), \tag{6.1.34}$$

with φ *elliptic of order* μ *on* Γ, g *satisfying* (6.1.27).

Consider the symbol $Q(x,\xi)$, defined to be the left side of (6.1.33). We have

$$Q(x,\xi) \in S^{2\mu}_{1,\delta} \cap C^{r-1}S^{2\mu}_{1,0}, \tag{6.1.35}$$

by (6.1.30), so we can apply the sharp Gårding inequality of §2.4, to obtain

$$\text{Re }(Q(x,D)v,v) \geq -C_1\|v\|^2_{L^2} \tag{6.1.36}$$

provided

$$\mu \leq (r-1)/(r+1) \text{ and } \mu \leq \rho/2. \tag{6.1.37}$$

In light of Lemma 6.1.B, we can replace $c\{c,a\}(x,D)$ in $Q(x,D)$ by Re $(1/i)C^*[C,A]$, and still have the estimate (6.1.36), provided also

$$\mu \leq 1 - \frac{\delta}{2} \quad (\text{if } r > 2). \tag{6.1.38}$$

However, this condition follows automatically from the second part of (6.1.37). If these conditions hold, we get a lower estimate on the left side of (6.1.18), yielding

(6.1.39)
$$\|\varphi_\varepsilon(x,D)v\|^2_{L^2} - \|E(x,D)v\|^2_{L^2} - C_1\|v\|^2_{L^2} \leq \|C_\varepsilon Pv\|^2_{L^2} + |(Wv,v)|.$$

Now, if W is given by (6.1.19), with $C = C_\varepsilon$ bounded in $OPS^\mu_{1,0}$, we have a priori that $W = W_\varepsilon$ is bounded in $OPS^{2\mu-(1-\delta)}_{1,\delta}$, given $B \in OPS^0_{1,\delta}$, but in fact, we can say more. From (6.1.5) it follows that, for the symbol $b(x,\xi)$ of B,

$$D^\alpha_x b(x,\xi) \in S^0_{1,\delta} \text{ for } |\alpha| \leq \ell - 1, \tag{6.1.40}$$

if $\ell < r < \ell + 1$. Thus

$$r > 2 \Longrightarrow W_\varepsilon \text{ bounded in } OPS^{2\mu-1}_{1,\delta}, \tag{6.1.41}$$

and hence

$$|(Wv, v)| \leq C_2 \|v\|^2_{H^{\mu-\frac{1}{2}}}. \tag{6.1.42}$$

Thus (6.1.39) yields the estimate

$$\|\varphi_\varepsilon(x, D)v\|^2_{L^2} \leq \|C_\varepsilon Pv\|^2_{L^2} + K\|v\|^2_{L^2}, \tag{6.1.43}$$

assuming $\mu - \frac{1}{2} \leq 0$. Provided $r > 2$, we can take $\mu = \frac{1}{3}$, $\delta \in [\frac{1}{3}, 1)$, and have (6.1.37) - (6.1.38) hold. The estimate (6.1.43) holds provided a priori that v belongs to L^2. Taking $\varepsilon \to 0$ then gives $v \in H^\mu$ microlocally along γ, provided that $Pv \in H^\mu$ microlocally along γ and that $v \in H^\mu$ microlocally on a conic neighborhood of the endpoint q. This exhibits the prototypical propagation of singularities phenomenon. Note that we can construct $\beta(x, \xi) \in S^0_{cl}$, supported on a small conic neighborhood of γ, equal to 1 on a smaller conic neighborhood, and then $v_1 = \beta(x, D)v$ belongs to H^μ, while Pv_1 coincides with Pv microlocally on a small conic neighborhood of γ.

Now this argument works if P is replaced by P_σ of (6.1.7), for any $\sigma \in \mathbb{R}$. The difference $P - P_\sigma$ satisfies (6.1.9), and its symbol also has the same property (6.1.40) that $b(x, \xi)$ does. Therefore, microlocal cut-offs and iterations of the argument above yield the following.

PROPOSITION 6.1.D. *Let $v \in \mathcal{D}'$ solve $M^\# v = g$. Let γ be as above. If g belongs to H^σ microlocally on γ and $v \in H^{m-1+\sigma}$ microlocally near q, then $v \in H^{m-1+\sigma}$ microlocally on γ.*

By (6.1.1) - (6.1.3), we can use this to establish our propagation of singularities result for solutions to nonlinear PDE. We have:

THEOREM 6.1.E. *Let $u \in C^{m+r}(\Omega)$, $r > 2$, solve the nonlinear PDE (6.1.1), and let γ be a null bicharacteristic curve of the linearized operator, $q \in \gamma$. Assume $f \in C^\infty$,*

$$u \in H^{m+\sigma} \tag{6.1.44}$$

on Ω, and

$$u \in H^{m-1+\sigma+s} \tag{6.1.45}$$

microlocally at q, with

$$0 \le s < r. \tag{6.1.46}$$

Then u satisfies (6.1.45) *microlocally on γ.*

It is clear that the assumption $r > 2$ can be weakened for restricted classes of operators $F(x, D^m u)$, particularly in the semilinear case; (6.1.44) can also be weakened in the semilinear case, since what is behind it is (6.1.3).

§6.2. Nonlinear formation of singularities

In [RR] there appear examples of the formation, in solutions to nonlinear wave equations, of extra singularities, arising from nonlinear interactions rather than propagated by a Hamiltonian flow. We will discuss a few simple examples here. Many other examples can be found in [Be2] and references given there. Our examples involve systems of equations. There are in the literature many examples of a similar nature for solutions to scalar equations, but generally you have to work harder to demonstrate their existence.

We begin with a simple 3×3 system in one space variable:

$$\begin{aligned} \partial_t u - \partial_x u &= 0 \\ \partial_t v + \partial_x v &= 0 \\ \partial_t w \qquad &= uv \end{aligned} \tag{6.2.1}$$

Take initial data

$$u(0,x) = H(1-x), \quad v(0,x) = H(1+x), \quad w(0,x) = 0, \tag{6.2.2}$$

where $H(s) = 1$ for $s \ge 0$, $H(s) = 0$ for $s < 0$. Clearly $u(t,x)$ and $v(t,x)$ are given by

$$u(t,x) = H(1-x-t), \quad v(t,x) = H(1+x-t), \tag{6.2.3}$$

and then w is obtained as

$$w(t,x) = \int_0^t u(s,x)v(s,x)ds. \tag{6.2.4}$$

Thus u is singular along the line $x + t = 1$ and v is singular along the line $x - t = -1$. For $t < 1$, w is singular along the union of these two lines; one would typically have this sort of singularity if uv were replaced by a linear function of u and v in (6.2.1). However, for $t > 1$, w is also singular along the line $x = 0$. Indeed, for $-1 \leq x \leq 1$, $t \geq 1$, we have explicitly

$$w(t,x) = 1 - |x|. \tag{6.2.5}$$

This is the extra singularity created by the nonlinear interaction at $t = 1$, $x = 0$.

The next example involves second order wave equations in two space variables. Thus, with $\Delta = \partial^2/\partial x_1^2 + \partial^2/\partial x_2^2$, let

$$\begin{aligned} \partial^2 u_j/\partial t^2 - \Delta u_j &= 0, \quad j = 1,2,3 \\ \partial^2 v/\partial t^2 - \Delta v &= u_1 u_2 u_3 \end{aligned} \tag{6.2.6}$$

and set initial conditions such that $u_j(t,x)$ are piecewise constant, jumping across a characteristic hyperplane Σ_j, taking the value 1 on one side Ω_j and the value 0 on the other side, in such a fashion that $\mathcal{P} = \Omega_1 \cap \Omega_2 \cap \Omega_3$ looks like an inverted 'pyramid,' with vertex at $p = (1,0,0) = (t, x_1, x_2)$, and such that $t \geq 1$ on $\mathcal{P}$. For v, set the initial condition $v(0,x) = v_t(0,x) = 0$, so $v = 0$ for $t < 1$. Note that there are constant coefficient vector fields X_j, parallel to the faces of $\mathcal{P}$, such that

$$L = X_1 X_2 X_3 \Longrightarrow L(u_1 u_2 u_3) = \delta_p$$

where δ_p is the point mass at the vertex p of $\mathcal{P}$. Hence $w = Lv$ satisfies

$$\partial^2 w/\partial t^2 - \Delta w = \delta_p, \quad w(t,x) = 0 \text{ for } t < 1. \tag{6.2.7}$$

Hence w is given by the fundamental solution to the wave equation, with a well known singularity on the forward light cone $\mathcal{C}_p$ emanating from p. Since applying L cannot *increase* the singular support, it follows that u in (6.2.6) also has this extra singularity along the cone $\mathcal{C}_p$.

The examples above can be modified to produce smoother (not C^∞) solutions, more directly illuminating the results of §6.1, which involve solutions with enough regularity to be continuous. It is also possible to extend results of §6.1 to apply to classes of discontinuous solutions, for classes of semilinear equations. In the quasilinear case, the phenomena become essentially different. For example, for 3×3 quasilinear equations in one space variable, if two shocks interact, the interaction often produces, not a third weaker singularity, but rather a third shock; cf [L2]. An analogue of this stronger sort of interaction arises in the rather subtle problem of interaction of oscillatory solutions to quasilinear equations; cf [Mj4] and references given there, and also [JMR].

In the examples given above it is clear that the mechanism behind the formation of extra singularities in the solutions to the PDE (6.2.1) and

(6.2.6) is the creation of extra singularities in taking nonlinear functions $F(u)$, i.e., $WF(F(u))$ can be bigger than $WF(u)$. Typically, elements in $WF(F(u))\setminus WF(u)$ intersected with the characteristic variety for the PDE are propagated. In light of the formula

$$F(u) = M_F(u; x, D)u \bmod C^\infty,$$

this illustrates the failure of $M_F(u; x, D) \in OPS^0_{1,1}$ to be microlocal, though it is pseudolocal, since Proposition 0.2.A continues to hold, which is consistent with the fact that $F(u)$ is C^∞ on any open set where u is C^∞.

§6.3. Egorov's theorem

We want to examine the behavior of operators obtained by conjugating a pseudodifferential operator P_0 by the solution operator to a 'hyperbolic' equation of the form

$$\partial u/\partial t = iA(t, x, D_x)u, \tag{6.3.1}$$

where we assume

$$A(t, x, \xi) \in {}^rS^1_{1,\delta} \tag{6.3.2}$$

is of the form $A = A_1 + A_2$ with A_1 *real* (scalar) and

$$A_0 \in {}^rS^0_{1,\delta}. \tag{6.3.3}$$

For simplicity we first look at the case where $A = A(x, D)$ does not depend on t. The conjugated operators we want to look at then are

$$P(t) = e^{itA}\ P_0\ e^{-itA}. \tag{6.3.4}$$

We will be able to get a rather precise analysis when δ in (6.3.2) is not too large, using a more elaborate version of the analysis given in §0.9. Then the $\delta = 1$ case will be analyzed via symbol smoothing.

Let X be the Hamiltonian field H_{A_1}; by (6.3.2) the coefficients of $\partial/\partial x_j$ belong to ${}^rS^0_{1,\delta}$ and those of $\partial/\partial \xi_j$ belong to ${}^{r-1}S^1_{1,\delta}$, if $r \ge 1$. Given (x_0, ξ_0), let Ω be the region $|x - x_0| \le 1$, $|\xi - \xi_0| \le \frac{1}{2}|\xi_0|$. Map this by the natural affine map to the 'standard' region R, defined by $|x| \le 1$, $|\xi| \le 1$. Let Y be the vector field on R corresponding to X; Y depends on (x_0, ξ_0). With $z = (x, \xi)$, we have

$$Y(z) = \sum_{j=1}^{2n} Y_j(z)\partial/\partial z_j \tag{6.3.5}$$

with

$$|D_z^\alpha Y_j(z)| \le C \qquad \text{for } |\alpha| \le r-1, \tag{6.3.6}$$
$$C_\alpha \; M^{|\alpha|-(r-1)}, \text{ for } |\alpha| \ge r-1,$$

where

$$M = |\xi_0|^\delta. \tag{6.3.7}$$

We will assume that $r \ge 2$, so there is a uniform bound on the first order derivatives of the coefficients of Y.

We want to examine the flow generated by Y :

$$dF/dt = Y(F), \quad F(0,z) = z. \tag{6.3.8}$$

We need to estimate the z-derivatives of F. Note that $F_1 = D_z F(z) v_1$ satisfies

$$dF_1/dt = DY(F)F_1, \quad F_1(0,z) = v_1. \tag{6.3.9}$$

This gives uniform bounds (independent of M) on $F_1 = D_z F$, for $|t| \le T_0$. Then $F_2 = D^2F(z)(v_1, v_2)$ satisfies

$$dF_2/dt = DY(F)F_2 + D^2Y(F)(F_1, v_2), \quad F_2(0,z) = 0. \tag{6.3.10}$$

Consequently

$$|D_z^2 F| \le C \; M^{(2-(r-1))_+}. \tag{6.3.11}$$

More generally,

$$|D_z^j F| \le C_j \; M^{(j-r+1)_+}. \tag{6.3.12}$$

Note that if we had $r < 2$, we would be saddled with the useless bound $|D_z F| \le C \; e^{TM}$.

With these estimates, we are in a position to prove the following. Let $\mathcal{C}_t$ be the flow generated by H_{A_1}.

PROPOSITION 6.3.A. *If (6.3.2)-(6.3.3) hold, with $r \ge 2$, then*

$$p(x,\xi) \in S^m_{\rho,\delta} \Longrightarrow p(\mathcal{C}_t(x,\xi)) \in S^m_{\rho,\delta}, \tag{6.3.13}$$

where $\rho = 1 - \delta$.

PROOF: We need estimates in each region Ω as described above. Let q be the associated function on R (q depends on (x_0, ξ_0)). The hypothesis on p implies

$$|D_z^\alpha q(z)| \le C_\alpha |\xi_0|^m \; M^{|\alpha|}; \tag{6.3.14}$$

recall $M = |\xi_0|^\delta$. We claim $q \circ F(t,z)$ has the same estimate. This follows from the chain rule and the estimates (6.3.12).

We now construct inductively the symbol of a pseudodifferential operator $Q(t)$, which we will show agrees with $P(t)$ in (6.3.4), for restricted δ, using an argument parallel to that in (0.9.7)- (0.9.10). We start with the assumption $\delta < 1$. We look for

$$Q(t,x,\xi) \sim Q_0(t,x,\xi) + Q_1(t,x,\xi) + \cdots, \tag{6.3.15}$$

beginning with

$$Q_0(t,x,\xi) = P_0(\mathcal{C}_t(x,\xi)) \in S^m_{\rho,\delta}, \tag{6.3.16}$$

by (6.3.13). We want to construct $Q(t)$ solving

$$Q'(t) = i[A(x,D), Q(t)] + R(t), \quad Q(0) = P_0, \tag{6.3.17}$$

where $R(t)$ is a family of smoothing operators. The symbol of $[A(x,D), Q(t)]$ is asymptotic to

$$H_{A_1}Q + \{A_0, Q\} + i\sum_{|\alpha|\geq 2}(1/\alpha!)\big[A^{(\alpha)}Q_{(\alpha)} - Q^{(\alpha)}A_{(\alpha)}\big], \tag{6.3.18}$$

where $A^{(\alpha)}(x,\xi) = D^\alpha_\xi(x,\xi)$, $A_{(\alpha)}(x,\xi) = i^{-|\alpha|}D^\alpha_x A(x,\xi)$. We want the difference of $\partial_t Q(t,x,\xi)$ and (6.3.18) to be asymptotic to zero. Thus we set up transport equations for $Q_j(t,x,\xi)$ in (6.3.15), $j \geq 0$, beginning with

$$\partial Q_0/\partial t - H_{A_1}Q_0 = 0, \quad Q_0(0,x,\xi) = P_0(x,\xi), \tag{6.3.19}$$

which is solved by (6.3.16). The transport equation for Q_1 will hence be

$$\begin{aligned}\partial Q_1/\partial t - H_{A_1}Q_1 &\sim \{A_0, Q_0\} + \sum_{|\alpha|\geq 2}(i/\alpha!)\big[A^{(\alpha)}Q_{0(\alpha)} - Q_0^{(\alpha)}A_{(\alpha)}\big] \\ &= B_1(t,x,\xi),\end{aligned} \tag{6.3.20}$$

with initial condition

$$Q_1(0,x,\xi) = 0. \tag{6.3.21}$$

It is necessary to examine $B_1(t,x,\xi)$. As long as (6.3.3) holds, or even more generally whenever $A_0 \in S^0_{1,\delta}$,

$$Q_0 \in S^m_{\rho,\delta} \Longrightarrow \{A_0, Q_0\} \in S^{m-(\rho-\delta)}_{\rho,\delta}. \tag{6.3.22}$$

For this to be useful we will need $\rho > \delta$, i.e., $\delta < \frac{1}{2}$. Continuing, we see that

$$A^{(\alpha)}Q_{0(\alpha)} \in S^{m+1-|\alpha|+\delta|\alpha|}_{\rho,\delta} \subset S^{m-1+2\delta}_{\rho,\delta} \quad \text{if } |\alpha| \geq 2, \tag{6.3.23}$$

while

$$\begin{aligned} A_{(\alpha)}Q_0^{(\alpha)} &\in S^{m+1+\delta(|\alpha|-2)-\rho|\alpha|}_{\rho,\delta} \\ &\subset S^{m+1-2\rho}_{\rho,\delta} \qquad \text{if } |\alpha| \geq 2, \end{aligned} \tag{6.3.24}$$

provided (6.3.2) holds with $r \geq 2$. We see that $B_1(t,x,\xi)$ has order strictly less than m provided $\delta < \frac{1}{2}$, so $\rho > \frac{1}{2}$.

Higher order transport equations are treated similarly, and we obtain

$$Q_j(t,x,\xi) \in S^{m-(2\rho-1)j}_{\rho,\delta} = S^{m-(\rho-\delta)j}_{\rho,\delta} \tag{6.3.25}$$

provided $\rho = 1-\delta > \frac{1}{2}$. Thus Q, given by (6.3.15), satisfies (6.3.17), with $R(t)$ smoothing. To complete the analysis of (6.3.4), we need to compare $Q(t)$ with $P(t)$. Consider

$$V(t) = [Q(t) - P(t)]e^{itA} = Q(t)e^{itA} - e^{itA}P_0. \tag{6.3.26}$$

We have

$$V'(t) = iAV(t) + R(t)e^{itA}, \quad V(0) = 0. \tag{6.3.27}$$

Thus, for any $u \in \mathcal{D}'$, $v(t) = V(t)u$ satisfies

$$\partial v/\partial t = iAv + g(t), \quad v(0) = 0, \tag{6.3.28}$$

where $g(t) = R(t)e^{itA}u$ is smooth. Hence standard energy estimates show $v(t)$ is smooth, so $V(t) \in OPS^{-\infty}$. We have proved the following Egorov-type theorem.

PROPOSITION 6.3.B. *If A satisfies (6.3.2)-(6.3.3) with $r \geq 2$ and $P_0 \in OPS^m_{\rho,\delta}$ with $\rho = 1-\delta > \frac{1}{2}$, then*

$$P(t) = e^{itA}\, P_0\, e^{-itA} = P(t,x,D) \in OPS^m_{\rho,\delta} \tag{6.3.29}$$

and

$$P(t,x,\xi) - P_0(\mathcal{C}_t(x,\xi)) \in S^{m-(\rho-\delta)}_{\rho,\delta}, \tag{6.3.30}$$

where $\mathcal{C}_t$ is the flow generated by H_{A_1}.

We now look at conjugates

$$P(t) = e^{itM}\, P_0\, e^{-itM}, \tag{6.3.31}$$

where

$$M(x,\xi) \in \mathcal{A}_0^r S^1_{1,1}, \ M = M_1 + M_2, \ M_1 \text{ real}, \ M_0 \in \mathcal{A}_0^r S^0_{1,1}. \tag{6.3.32}$$

Using the symbol smoothing of §1.3, write

$$M(x,\xi) = A(x,\xi) + B(x,\xi) \tag{6.3.33}$$

where

$$A(x,\xi) \in \mathcal{A}_0^r S^1_{1,\delta}, \quad B(x,\xi) \in S^{1-r\delta}_{1,1}. \tag{6.3.34}$$

The symbol $A(x,\xi)$ satisfies (6.3.2)-(6.3.3). Provided $r > 2$, we can take δ slightly less than $\frac{1}{2}$ so that

$$B(x,\xi) \in S^{-\gamma}_{1,1}, \quad \gamma = r\delta - 1 > 0. \tag{6.3.35}$$

We can apply Proposition 6.3.B (with a slight change of notation), analyzing

$$Q(t) = e^{itA} \ P_0 \ e^{-itA} = Q(t,x,D) \in OPS^0_{\rho,\delta}, \tag{6.3.36}$$

given $P_0 \in OPS^0_{\rho,\delta}$. We now want to compare the operators (6.3.31) and (6.3.36). Note that

$$Q'(t) = i[A,Q(t)], \quad P'(t) = i[A,P(t)] + R(t) \tag{6.3.37}$$

with

$$R(t) = i[B,P(t)] : H^s \longrightarrow H^{s+\gamma}, \quad \text{for } s > -\gamma. \tag{6.3.38}$$

Consequently, if we set

$$V(t) = [Q(t) - P(t)]e^{itA},$$

this time we have

$$V'(t) = iAV(t) - R(t)e^{itA}, \quad V(0) = 0.$$

Hence

$$V(t) : H^s \longrightarrow H^{s+\gamma}.$$

Therefore

$$e^{itM} \ P_0 \ e^{-itM} - e^{itA} \ P_0 \ e^{-itA} : H^s \longrightarrow H^{s+\gamma}, \text{ for } s > -\gamma. \tag{6.3.39}$$

It is possible to apply this result to study propagation of singularities, in analogy with the proof of Proposition 0.10.E. Indeed, suppose $u \in H^\sigma_{mcl}(\Gamma)$, so if $P_0 \in OPS^0$ has symbol suported inside Γ, $P_0 u \in H^\sigma$. If A satisfies the hypotheses of Proposition 6.3.B, $e^{itA}u$ is mapped to H^σ by $P(t)$ in (6.3.29). Thus, if $\mathcal{C}_t\Gamma$ contains a cone Γ_t, we deduce that $e^{itA}u \in H^\sigma_{mcl}(\Gamma_t)$. We also claim that, if M satisfies (6.3.32) with $r > 2$,

$$(e^{itM} P_0 e^{-itM})(e^{itM}u) \in H^\sigma \text{ if } \sigma > -\gamma. \tag{6.3.40}$$

This follows from:

LEMMA 6.3.C. *If $M(x,\xi)$ satisfies (6.3.32) with $r > 2$, then*

$$e^{itM} : H^s \longrightarrow H^s, \quad s > -\gamma. \tag{6.3.41}$$

PROOF: Given $u \in H^s$, $e^{itM}u = v(t)$ satisfies

$$\partial v/\partial t = iAv + iBv, \quad v(0) = u. \tag{6.3.42}$$

As long as (6.3.35) holds with $\gamma > 0$, standard linear hyperbolic techniques apply to (6.3.42), yielding (6.3.41).

Given (6.3.40), we can use (6.3.39) to conclude

$$P(t)(e^{itM}u) \in H^\sigma \text{ if } s \le \sigma \le s + \gamma \tag{6.3.43}$$

provided

$$u \in H^s, \quad s > -\gamma. \tag{6.3.44}$$

Thus, under these hypotheses,

$$e^{itM}u \in H^\sigma_{mcl}(\Gamma_t). \tag{6.3.45}$$

This yields a propagation of singularities result along the lines of Theorem 6.1.E, but substantially weaker, so we will not write out the details.

Chapter 7: Nonlinear parabolic systems

We examine existence, uniqueness, and regularity of solutions to nonlinear parabolic systems. We begin with an approach to strongly parabolic quasilinear equations using techniques very similar to those applied to hyperbolic systems in Chapter 5, moving on to symmetrizable quasilinear parabolic systems in §7.2. It turns out that another approach, making stronger use of techniques of Chapter 3, yields sharper results. We explore this in §7.3, treating there completely nonlinear as well as quasilinear systems. For a class of scalar equations in divergence form, we make contact with the DeGiorgi-Nash-Moser theory and show how some global existence results follow.

In §7.4 we consider semilinear parabolic systems. We first state a result which is just a specialization of Proposition 7.3.C, which applies to a class of quasilinear equations. Then, by a more elementary method, we derive a result for initial data in $C^1(M)$. Neither of these two results contains the other, so having them both may yield useful information. As one important example of a semilinear parabolic system, we consider the parabolic equation approach to existence of harmonic maps $M \to N$, when N has negative sectional curvature, due to Eells and Sampson. In outline our analysis follows that presented in [J], with some simplifications arising from taking N to be imbedded in $\mathbb{R}^k$ (as in [Str]), and also some simplifications in the use of parabolic theory.

§7.1. Strongly parabolic quasilinear systems

In this section we study the initial value problem

$$\partial u/\partial t = \sum_{j,k} A^{jk}(t,x,D_x^1 u)\partial_j\partial_k u + B(t,x,D_x^1 u), \quad u(0) = f. \tag{7.1.1}$$

Here, u takes values in $\mathbb{R}^K$, and each A^{jk} can be a symmetric $K \times K$ matrix; we assume A^{jk} and B are smooth in their arguments. As in Chapter 5 we assume for simplicity that $x \in M$, an n-dimensional torus. The strong parabolicity condition we impose is

$$\sum_{j,k} A^{jk}(t,x,D_x^1 u)\xi_j\xi_k \geq C_0|\xi|^2 I. \tag{7.1.2}$$

The analysis will be in many respects similar to that in §5.1. We consider the approximating equation

$$\begin{aligned}\partial u_\varepsilon/\partial t &= J_\varepsilon \sum A^{jk}(t,x,D_x^1 J_\varepsilon u_\varepsilon)\partial_j\partial_k J_\varepsilon u_\varepsilon + J_\varepsilon B(t,x,D_x^1 J_\varepsilon u_\varepsilon) \\ &= J_\varepsilon L_\varepsilon J_\varepsilon u_\varepsilon + B_\varepsilon,\end{aligned} \tag{7.1.3}$$

which, for any fixed $\varepsilon > 0$, has a solution satisfying $u_\varepsilon(0) = f$. To estimate $\|u_\varepsilon(t)\|_{H^s}$, we consider

$$\frac{d}{dt}(\Lambda^s u_\varepsilon, \Lambda^s u_\varepsilon) = 2(\Lambda^s J_\varepsilon L_\varepsilon J_\varepsilon u_\varepsilon, \Lambda^s u_\varepsilon) + 2(\Lambda^s B_\varepsilon, \Lambda^s u_\varepsilon). \tag{7.1.4}$$

The last term in (7.1.4) is easy. We have

$$(\Lambda^s B_\varepsilon, \Lambda^s u_\varepsilon) \le C(\|J_\varepsilon u_\varepsilon\|_{C^1})\|J_\varepsilon u_\varepsilon\|_{H^{s+1}} \cdot \|J_\varepsilon u_\varepsilon\|_{H^s}. \tag{7.1.5}$$

To analyze the first term on the right side of (7.1.4), write it as 2 times

$$(\Lambda^s L_\varepsilon J_\varepsilon u_\varepsilon, \Lambda^s J_\varepsilon u_\varepsilon) = (L_\varepsilon \Lambda^s J_\varepsilon u_\varepsilon, \Lambda^s J_\varepsilon u_\varepsilon) + ([\Lambda^s, L_\varepsilon] J_\varepsilon u_\varepsilon, \Lambda^s J_\varepsilon u_\varepsilon). \tag{7.1.6}$$

Applying the Kato-Ponce estimate to

$$[\Lambda^s, L_\varepsilon] = \sum [\Lambda^s, A^{jk}(t,x,D_x^1 J_\varepsilon u_\varepsilon)]\partial_j \partial_k, \tag{7.1.7}$$

we dominate the last term in (7.1.6) by

$$C(\|J_\varepsilon u_\varepsilon\|_{C^2})\|J_\varepsilon u_\varepsilon\|_{H^{s+1}} \cdot \|J_\varepsilon u_\varepsilon\|_{H^s}. \tag{7.1.8}$$

Now, to analyze the first term on the right side of (7.1.6), write

$$L_\varepsilon = \sum \partial_j A^{jk}(t,x,D_x^1 J_\varepsilon u_\varepsilon)\partial_k + \sum [A^{jk}(t,x,D_x^1 J_\varepsilon u_\varepsilon), \partial_j]\partial_k. \tag{7.1.9}$$

The contribution of the last term of (7.1.9) to the first term on the right side of (7.1.6) is also seen to be dominated by (7.1.8). Finally, by hypothesis (7.1.2), we have

$$-\sum \operatorname{Re}\,(\partial_j A^{jk} \partial_k v, v) \ge C_0 \|\nabla v\|_{L^2}^2. \tag{7.1.10}$$

Putting these estimates together yields for (7.1.4) the estimate

$$\begin{aligned} \frac{d}{dt}(\Lambda^s u_\varepsilon, \Lambda^s u_\varepsilon) &\le -C_1 \|J_\varepsilon u_\varepsilon\|_{H^{s+1}}^2 + C(\|J_\varepsilon u_\varepsilon\|_{C^2})\|J_\varepsilon u_\varepsilon\|_{H^{s+1}} \cdot \|J_\varepsilon u_\varepsilon\|_{H^s} \\ &\le -\frac{1}{2} C_1 \|J_\varepsilon u_\varepsilon\|_{H^{s+1}}^2 + C'(\|J_\varepsilon u_\varepsilon\|_{C^2})\|J_\varepsilon u_\varepsilon\|_{H^s}^2. \end{aligned} \tag{7.1.11}$$

From here, standard arguments entering the proofs of Proposition 5.1.B, Proposition 5.1.C, and Proposition 5.1.E apply, to yield:

PROPOSITION 7.1.A. *Given the strong parabolicity hypothesis* (7.1.2), *if* $f \in H^s(M)$ *and* $s > \frac{1}{2}n + 2$, *then* (7.1.1) *has a unique solution*

$$u \in L^\infty(I, H^s(M)) \cap \mathit{Lip}(I, H^{s-2}(M)) \tag{7.1.12}$$

for some interval $I = [0, T), T > 0$. *The solution persists as long as* $\|u(t)\|_{C^2}$ *is bounded.*

The argument establishing $u \in C(I, H^s(M))$ in Proposition 5.1.D does not quite yield that result right away in this setting, since it used reversiblity, valid for hyperbolic PDE but not for parabolic PDE. In this context, that argument does yield right continuity:

$$t_j \searrow t \text{ in } I \Longrightarrow u(t_j) \to u(t) \text{ in } H^s\text{-norm}. \tag{7.1.13}$$

In particular, we have continuity at $t = 0$. In fact, we have

$$u \in C(I, H^s(M)), \tag{7.1.14}$$

continuity at points $t \in (0, T)$ following from higher regularity, which we now establish.

In fact, for any $S < T$, if we integrate (7.1.11) over $J = [0, S)$, we obtain a bound on $\int_J \|J_\varepsilon u_\varepsilon(\tau)\|^2_{H^{s+1}} d\tau$, if $s > \frac{1}{2}n + 2$, so that $\|u\|_{C^2} \leq C\|u\|_{H^s}$ and $\|J_\varepsilon u_\varepsilon(t)\|_{H^s}$ gets bounded. Passing to the limit $\varepsilon \to 0$, we have

$$u \in L^2(J, H^{s+1}(M)). \tag{7.1.15}$$

This implies that $u(t_1)$ 'exists' in $H^{s+1}(M)$ for *almost all* $t_1 \in J$. We expect that $u(t)$ coincides with the solution v to (7.1.1), with $v(t_1) = u(t_1)$, for $t > t_1$, implying we can replace s by $s + 1$ in (7.1.12) and (7.1.14), at least on (t_1, T). Iterating this heuristic argument leads to the following regularity result, for which we provide a rigorous proof.

PROPOSITION 7.1.B. *The solution* u *of Proposition* 7.1.A *has the property*

$$u \in C^\infty((0, T) \times M). \tag{7.1.16}$$

PROOF: Fix any $S < T$ and take $J = [0, S)$. Passing to a subsequence, we can suppose that, with $v_j(t) = J_{\varepsilon_j} u_{\varepsilon_j}(t)$,

$$\|v_{j+1} - v_j\|_{L^2(J, H^{s+1}(M))} \leq 2^{-j}. \tag{7.1.17}$$

Thus, if we consider

$$(7.1.18)\quad \Phi(t) = \sup_j \|v_j(t)\|_{H^{s+1}} \le \|v_1(t)\|_{H^{s+1}} + \sum_{j=1}^{\infty} \|v_{j+1}(t) - v_j(t)\|_{H^{s+1}},$$

we deduce that $\Phi \in L^2(J)$, and in particular $\Phi(t) < \infty$ almost everywhere. Let
$\mathcal{T} = \{t \in J : \Phi(t) < \infty\}$. Thus, for any $t_1 \in \mathcal{T}$, $\{J_{\varepsilon_j} u_{\varepsilon_j}(t_1) : j \ge 1\}$ is bounded in $H^{s+1}(M)$. It converges to $u(t_1)$ in $H^{s-2}(M)$; hence $u(t_1) \in H^{s+1}(M)$. By uniqueness, the solution v to (7.1.1), $v(t_1) = u(t_1)$, which belongs to $C([t_1, S), H^{s+1}(M))$, coincides with $u(t)$ on $[t_1, S)$. We can iterate this argument, to get $u \in C(I, C^\infty(M))$, and then (7.1.16) easily follows.

It is of interest to look at the following special case of (7.1.1),

$$(7.1.19)\qquad \partial u/\partial t = \sum A^{jk}(t,x,u)\partial_j\partial_k u + B(t,x,D_x^1 u), \quad u(0) = f,$$

in which the coefficients A^{jk} depend on u but not its derivatives. Consequently the bound (7.1.8) can be improved, replacing $C(\|J_\varepsilon u_\varepsilon\|_{C^2})$ by $C(\|J_\varepsilon u_\varepsilon\|_{C^1})$, and this leads to a corresponding improvement in (7.1.11). Hence we have:

PROPOSITION 7.1.C. *If (7.1.19) is strongly parabolic, and if $f \in H^s(M)$, $s > \frac{1}{2}n + 1$, then there is a unique solution*

$$u \in C([0,T), H^s(M)) \cap C^\infty((0,T) \times M),$$

which persists as long as $\|u(t)\|_{C^1}$ is bounded.

§7.2. Petrowski parabolic quasilinear equations

We continue to study the initial value problem (7.1.1), but we replace the strong parabolicity hypothesis (7.1.2) with the following more general hypothesis on

$$(7.2.1)\qquad L_2(t,v,x,\xi) = -\sum_{j,k} A^{jk}(t,x,v)\xi_j\xi_k,$$

namely

$$(7.2.2)\qquad \operatorname{spec} L_2(t,v,x,\xi) \subset \{z \in \mathbb{C} : \operatorname{Re} z \le -C_0|\xi|^2\}$$

for some $C_0 > 0$. Again we will try to produce the solution to (7.1.1) as a limit of solutions u_ε to (7.1.3). In order to get estimates, we construct a symmetrizer.

LEMMA 7.2.A. *Granted (7.2.1), there exists $P_0(t,v,x,\xi)$, smooth in its arguments, for $\xi \neq 0$, homogeneous of degree 0 in ξ, positive definite (i.e., $P_0 \geq cI > 0$), such that $-(P_0L_2 + L_2^*P_0)$ is also positive definite, i.e.,*

$$-(P_0L_2 + L_2^*P_0) \geq C|\xi|^2 I > 0. \tag{7.2.3}$$

The symmetrizer P_0, which is not unique, is constructed by establishing first that if L_2 is a *fixed* $K \times K$ matrix with spectrum in Re $z < 0$, then there exists a $K \times K$ matrix P_0 such that P_0 and $-(P_0L_2+L_2^*P_0)$ are positive definite. This is an exercise in linear algebra. One then observes the following facts. One, for a given positive matrix P_0, the set of L_2 such that $-(P_0L_2 + L_2^*P_0)$ is positive definite, is open. Next, for given L_2 with spectrum in Re $z < 0$, the set $\{P_0 : P_0 > 0, -(P_0L_2 + L_2^*P_0) > 0\}$, is an open convex set of matrices, within the linear space of self adjoint $K \times K$ matrices. Using this and a partition of unity argument, one can establish the following, which then yields Lemma 7.2.A.

LEMMA 7.2.B. *If $\mathcal{M}_K^-$ denotes the space of real $K \times K$ matrices with spectrum in Re $z < 0$ and $\mathcal{P}_K^+$ the space of positive definite (complex) $K \times K$ matrices, there is a smooth map*

$$\Phi : \mathcal{M}_K^- \longrightarrow \mathcal{P}_K^+,$$

homogeneous of degree 0, such that, if $L \in \mathcal{M}_K^-$ and $P = \Phi(L)$, then $-(PL + L^*P) \in \mathcal{P}_K^+$.

Having constructed $P_0(t,v,x,\xi)$, note that

$$\begin{aligned} u \in C^2 \Longrightarrow L(t, D_x^1 u, x, \xi) &\in C^1 S_{cl}^2 \text{ and} \\ P_0(t, D_x^1 u, x, \xi) &\in C^1 S_{cl}^0. \end{aligned} \tag{7.2.4}$$

Now, with $P = P_0(t, D_x^1 u, x, D)$, set

$$Q = \frac{1}{2}(P + P^*) + K\Lambda^{-1}, \tag{7.2.5}$$

with $K > 0$ chosen so that Q is positive definite on L^2. Now, with u_ε defined as the solution to (7.1.3), $u_\varepsilon(0) = f$, we estimate

$$\frac{d}{dt}(\Lambda^s u_\varepsilon, Q_\varepsilon \Lambda^s u_\varepsilon) = 2(\Lambda^s(\partial u_\varepsilon/\partial t), Q_\varepsilon \Lambda^s u_\varepsilon) + (\Lambda^s u_\varepsilon, Q'_\varepsilon \Lambda^s u_\varepsilon), \tag{7.2.6}$$

where Q_ε is obtained as in (7.2.5) from $P_\varepsilon = P_0(t, D_x^1 J_\varepsilon u_\varepsilon, x, D)$. In the last term we can replace Q'_ε by $(d/dt)P_0(t, D_x^1 J_\varepsilon u_\varepsilon, x, D)$, and obtain

$$|(\Lambda^s u_\varepsilon, Q'_\varepsilon \Lambda^s u_\varepsilon)| \leq C(\|u_\varepsilon(t)\|_{C^3})\|u_\varepsilon(t)\|_{H^s}^2. \tag{7.2.7}$$

The C^3-norm arises from the equation (7.1.3) for $\partial u_\varepsilon/\partial t$.

We can write the first term on the right side of (7.2.6) as twice

$$(Q_\varepsilon \Lambda^s J_\varepsilon L_\varepsilon J_\varepsilon u_\varepsilon, \Lambda^s u_\varepsilon) + (Q_\varepsilon \Lambda^s B_\varepsilon, \Lambda^s u_\varepsilon), \tag{7.2.8}$$

where L_ε is as in (7.1.3). The last term here is easily dominated by

$$C(\|u_\varepsilon(t)\|_{C^1})\|J_\varepsilon u_\varepsilon(t)\|_{H^{s+1}} \cdot \|u_\varepsilon(t)\|_{H^s}. \tag{7.2.9}$$

We write the first term in (7.2.8) as

$$\begin{aligned}(Q_\varepsilon L_\varepsilon \Lambda^s J_\varepsilon u_\varepsilon, \Lambda^s J_\varepsilon u_\varepsilon) + (Q_\varepsilon[\Lambda^s, L_\varepsilon]J_\varepsilon u_\varepsilon, \Lambda^s J_\varepsilon u_\varepsilon)& \\ + ([Q_\varepsilon \Lambda^s, J_\varepsilon]L_\varepsilon J_\varepsilon u_\varepsilon, \Lambda^s u_\varepsilon)&,\end{aligned} \tag{7.2.10}$$

just as in (5.2.36), except now L_ε is a second order operator. As long as (7.2.4) holds, P_ε also has symbol in $C^1 S^0_{cl}$, and as in (5.2.37) we can apply Proposition 4.1.E to get

$$[Q_\varepsilon \Lambda^s, J_\varepsilon] \text{ bounded in } \mathcal{L}(H^{s-1}, L^2), \tag{7.2.11}$$

with a bound given in terms of $\|u_\varepsilon\|_{C^2}$. Furthermore, we have

$$\|L_\varepsilon J_\varepsilon u_\varepsilon\|_{H^{s-1}} \le C(\|u_\varepsilon\|_{C^1})\|J_\varepsilon u_\varepsilon\|_{H^{s+1}} + C(\|u_\varepsilon\|_{C^2})\|J_\varepsilon u_\varepsilon\|_{H^s}, \tag{7.2.12}$$

so we can dominate the last term in (7.2.10) by

$$C(\|u_\varepsilon(t)\|_{C^2})\|J_\varepsilon u_\varepsilon\|_{H^{s+1}} \cdot \|u_\varepsilon\|_{H^s}. \tag{7.2.13}$$

Moving to the second term in (7.2.10), we can use the Kato-Ponce estimate to get

$$\|[\Lambda^s, L_\varepsilon]v\|_{L^2} \le C\sum_{j,k}\Big[\|A^{jk}\|_{Lip^1} \cdot \|v\|_{H^{s+1}} + \|A^{jk}\|_{H^s} \cdot \|v\|_{C^2}\Big]. \tag{7.2.14}$$

Hence the second term in (7.2.10) is also bounded by (7.2.13).

This brings us to the first term in (7.2.10), and for this we apply the Gårding inequality, Proposition 4.3.A, to get

$$(Q_\varepsilon L_\varepsilon v, v) \le -C_0\|v\|^2_{H^1} + C(\|u_\varepsilon\|_{C^2})\|v\|^2_{L^2}. \tag{7.2.15}$$

Substituting $v = \Lambda^s J_\varepsilon u_\varepsilon$ and using the other estimates on terms from (7.2.6), we have

$$\begin{aligned}\frac{d}{dt}(\Lambda^s u_\varepsilon, Q_\varepsilon \Lambda^s u_\varepsilon) \le -C_0\|J_\varepsilon u_\varepsilon\|^2_{H^{s+1}}& \\ + C(\|u_\varepsilon\|_{C^3})\|u_\varepsilon\|_{H^s}\big[\|J_\varepsilon u_\varepsilon\|_{H^{s+1}} + \|u_\varepsilon\|_{H^s}\big]&\end{aligned}$$

which we can further dominate as in (7.1.11).

From here, all the other arguments yielding Proposition 7.1.A and Proposition 7.1.B apply, and we have the following.

PROPOSITION 7.2.C. *Given the parabolicity hypothesis* (7.2.1), *if* $f \in H^s(M)$, *and* $s > \frac{1}{2}n + 3$, *then* (7.1.1) *has a unique solution*

$$u \in C([0,T), H^s(M)) \cap C^\infty((0,T) \times M) \tag{7.2.16}$$

for some $T > 0$, *which persists as long as* $\|u(t)\|_{C^3}$ *is bounded.*

§7.3. Sharper estimates

While on the face of it §7.1 seems to be a clean parallel with the analysis of hyperbolic equations in Chapter 5, in fact the results are not as sharp as they can be, and we obtain sharper results here, making more use of paradifferential operator calculus. We begin with completely nonlinear equations:

$$\partial u/\partial t = F(t,x,D_x^2 u), \quad u(0) = f, \tag{7.3.1}$$

for u taking values in $\mathbb{R}^K$. We suppose $F = F(t,x,\zeta)$, $\zeta = (\zeta_{\alpha j} : |\alpha| \leq 2, 1 \leq j \leq K)$ is smooth in its arguments, and our strong parabolicity hypothesis is:

$$-\operatorname{Re} \sum_{|\alpha|=2} (\partial F/\partial \zeta_\alpha)\xi^\alpha \geq C|\xi|^2 I, \tag{7.3.2}$$

for $\xi \in \mathbb{R}^n$, where $\operatorname{Re} A = \frac{1}{2}(A + A^*)$ for a $K \times K$ matrix A. Using §3.3, we write

$$F(t,x,D_x^2 v) = M(v;t,x,D)v + R(v). \tag{7.3.3}$$

Thus, for $r > 0$,

$$v(t) \in C^{2+r} \Longrightarrow M(v;t,x,\xi) \in \mathcal{A}_0^r S_{1,1}^2 \subset C^r S_{1,0}^2 \cap S_{1,1}^2. \tag{7.3.4}$$

The hypothesis (7.3.2) implies

$$-\operatorname{Re} M(v;t,x,\xi) \geq C|\xi|^2 I > 0 \tag{7.3.5}$$

for $|\xi|$ large. Note that symbol smoothing in x gives

$$M(v;t,x,\xi) = M^\#(t,x,\xi) + M^b(t,x,\xi) \tag{7.3.6}$$

and, when (7.3.4) holds (for fixed t)

$$M^{\#}(t,x,\xi) \in \mathcal{A}_0^r S^2_{1,\delta}, \quad M^b(t,x,\xi) \in S^{2-r\delta}_{1,1}. \tag{7.3.7}$$

We also have

$$-\operatorname{Re} M^{\#}(t,x,\xi) \geq C|\xi|^2 I > 0 \tag{7.3.8}$$

for $|\xi|$ large.

We will obtain a solution to (7.3.1) as a limit of solutions u_ε to

$$\partial u_\varepsilon/\partial t = J_\varepsilon F(t,x,D_x^2 J_\varepsilon u_\varepsilon), \quad u_\varepsilon(0) = f. \tag{7.3.9}$$

Thus we need to show that $u_\varepsilon(t,x)$ exists on an interval $t \in [0,T)$ independent of $\varepsilon \in (0,1]$ and has a limit as $\varepsilon \to 0$ solving (7.3.1). As before, all this follows from an estimate on the H^s-norm, and we begin with

$$\begin{aligned} \frac{d}{dt}\|\Lambda^s u_\varepsilon(t)\|^2_{L^2} &= 2(\Lambda^s J_\varepsilon F(t,x,D_x^2 J_\varepsilon u_\varepsilon), \Lambda^s u_\varepsilon) \\ &= (\Lambda^s M_\varepsilon J_\varepsilon u_\varepsilon, \Lambda^s J_\varepsilon u_\varepsilon) + 2(\Lambda^s R_\varepsilon, \Lambda^s J_\varepsilon u_\varepsilon). \end{aligned} \tag{7.3.10}$$

The last term is easily bounded by

$$C(\|u_\varepsilon(t)\|_{L^2})\big[\|J_\varepsilon u_\varepsilon(t)\|^2_{H^s} + 1\big].$$

Here $M_\varepsilon = M(J_\varepsilon u_\varepsilon; t,x,D)$. Writing $M_\varepsilon = M_\varepsilon^{\#} + M_\varepsilon^b$ as in (7.3.6), we see that

$$\begin{aligned} &(\Lambda^s M_\varepsilon^b J_\varepsilon u_\varepsilon, \Lambda^s J_\varepsilon u_\varepsilon) \\ &\qquad = (\Lambda^{s-1} M_\varepsilon^b J_\varepsilon u_\varepsilon, \Lambda^{s+1} J_\varepsilon u_\varepsilon) \\ &\qquad \leq C(\|J_\varepsilon u_\varepsilon\|_{C^{2+r}})\|J_\varepsilon u_\varepsilon\|_{H^{s+1-r\delta}}\|J_\varepsilon u_\varepsilon\|_{H^{s+1}} \end{aligned} \tag{7.3.11}$$

for $s > 1$, since by (7.3.7), $M_\varepsilon^b : H^{s+1-r\delta} \longrightarrow H^{s-1}$. We next estimate

$$\begin{aligned} &(\Lambda^s M_\varepsilon^{\#} J_\varepsilon u_\varepsilon, \Lambda^s J_\varepsilon u_\varepsilon) \\ &\qquad = (M_\varepsilon^{\#} \Lambda^s J_\varepsilon u_\varepsilon, \Lambda^s J_\varepsilon u_\varepsilon) + ([\Lambda^s, M_\varepsilon^{\#}] J_\varepsilon u_\varepsilon, \Lambda^s J_\varepsilon u_\varepsilon). \end{aligned} \tag{7.3.12}$$

By (7.3.7), $[\Lambda^s, M_\varepsilon^{\#}] \in OPS^{s+2-r}_{1,\delta}$, if $0 < r < 1$, so the last term in (7.3.12) is bounded by

$$\begin{aligned} &(\Lambda^{-1}[\Lambda^s, M_\varepsilon^{\#}] J_\varepsilon u_\varepsilon, \Lambda^{s+1} J_\varepsilon u_\varepsilon) \\ &\qquad \leq C(\|u_\varepsilon\|_{C^{2+r}})\|J_\varepsilon u_\varepsilon\|_{H^{s+1-r}} \cdot \|J_\varepsilon u_\varepsilon\|_{H^{s+1}}. \end{aligned} \tag{7.3.13}$$

Finally, Gårding's inequality applies to $M_\varepsilon^\#$:

$$(M_\varepsilon^\# w, w) \leq -C_0 \|w\|_{H^1}^2 + C_1(\|u_\varepsilon\|_{C^{2+r}})\|w\|_{L^2}^2. \tag{7.3.14}$$

Putting together the previous estimates, we obtain

$$\frac{d}{dt}\|u_\varepsilon(t)\|_{H^s}^2 \leq -\frac{1}{2}C_0\|J_\varepsilon u_\varepsilon\|_{H^{s+1}}^2 + C(\|u_\varepsilon\|_{C^{2+r}})\|J_\varepsilon u_\varepsilon\|_{H^{s+1-r\delta}}^2, \tag{7.3.15}$$

and using Poincare's inequality, we can replace $-\frac{1}{2}C_0$ by $-\frac{1}{4}C_0$ and the $H^{s+1-r\delta}$-norm by the H^s-norm, getting

$$\begin{aligned}\frac{d}{dt}\|u_\varepsilon(t)\|_{H^s}^2 \leq -\frac{1}{4}C_0\|J_\varepsilon u_\varepsilon(t)\|_{H^{s+1}}^2 \\ + C'(\|u_\varepsilon(t)\|_{C^{2+r}})\|J_\varepsilon u_\varepsilon(t)\|_{H^s}^2.\end{aligned} \tag{7.3.16}$$

From here, the arguments used to establish Proposition 7.1.A and Proposition 7.1.B yield:

PROPOSITION 7.3.A. *If* (7.3.1) *is strongly parabolic and* $f \in H^s(M), s > \frac{1}{2}n + 2$, *then there is a unique solution*

$$u \in C([0,T), H^s(M)) \cap C^\infty((0,T) \times M), \tag{7.3.17}$$

which persists as long as $\|u(t)\|_{C^{2+r}}$ *is bounded, given* $r > 0$.

Note that if the method of quasilinearization were applied to (7.3.1) in concert with the results of §7.1, we would require $s > \frac{1}{2}n + 3$ and for persistence of the solution would need a bound on $\|u(t)\|_{C^3}$.

We take another look at the quasilinear case (7.1.1), i.e., the special case of (7.3.1) in which

$$F(t,x,D_x^2 u) = \sum A^{jk}(t,x,D_x^1 u)\partial_j\partial_k u + B(t,x,D_x^1 u). \tag{7.3.18}$$

We form $M(v;t,x,D)$ as before, by (7.3.3). In this case, we can replace (7.3.4) by

$$v \in C^{1+r} \Longrightarrow M(v;t,x,\xi) \in \mathcal{A}_0^r S_{1,1}^2 + S_{1,1}^{2-r}. \tag{7.3.19}$$

Thus we can produce a decomposition (7.3.6) such that (7.3.7) holds for $v \in C^{1+r}$. Hence the estimates (7.3.11)-(7.3.16) all hold with constants depending on the C^{1+r}-norm of $u_\varepsilon(t)$, rather than the C^{2+r}-norm, and we have the following improvement of Proposition 7.1.A-Proposition 7.1.B.

PROPOSITION 7.3.B. *If the quasilinear system* (7.1.1) *is strongly parabolic and* $f \in H^s(M), s > \frac{1}{2}n + 1$, *then there is a unique solution satisfying* (7.3.17), *which persists as long as* $\|u(t)\|_{C^{1+r}}$ *is bounded, given* $r > 0$.

We look at the further special case (7.1.19), where

$$F(t, x, D_x^2 u) = \sum A^{jk}(t, x, u)\partial_j \partial_k u + B(t, x, D_x^1 u). \tag{7.3.20}$$

In this case, if $r > 0$, we have

$$v \in C^r \Longrightarrow M(v; t, x, \xi) \in \mathcal{A}_0^r S_{1,1}^2 + S_{1,1}^{2-r}, \tag{7.3.22}$$

and the following improvement of Proposition 7.1.C results.

PROPOSITION 7.3.C. *If the system* (7.1.19) *is strongly parabolic and* $f \in H^s(M), s > \frac{1}{2}n$ *(and* $s > 1$ *if* $n = 1$*), then there is a unique solution satisfying* (7.3.17), *which persists as long as* $\|u(t)\|_{C^r}$ *is bounded, given* $r > 0$.

Let us note that the class of equations (7.1.19) is equivalent to the class of equations

$$\partial u/\partial t = \sum \partial_j A^{jk}(t, x, u)\partial_k u + B'(t, x, D_x^1 u), \quad u(0) = f. \tag{7.3.23}$$

We specialize to the case

$$\partial u/\partial t = \sum \partial_j A^{jk}(t, x, u)\partial_k u, \quad u(0) = f. \tag{7.3.24}$$

Then the DeGiorgi-Nash-Moser theory has the following implication, when the coefficients A^{jk} are scalar.

THEOREM 7.3.D. *Suppose* (7.3.24) *holds on* $[t_0, t_0 + a] \times M$, *with scalar coefficients satisfying*

$$\lambda_0 |\xi|^2 \leq \sum A^{jk}(t, x, u)\xi_j \xi_k \leq \lambda_1 |\xi|^2.$$

Then $u(t_0 + a, x) = w(x)$ *belongs to* C^r *for some* $r > 0$, *and there is an estimate*

$$\|w\|_{C^r} \leq K(a, \lambda_0, \lambda_1)\|u(t_0, \cdot)\|_{L^\infty}. \tag{7.3.25}$$

In particular, the factor $K(a, \lambda_0, \lambda_1)$ *does not depend on the modulus of continuity of* A^{jk}.

This produces a global existence result; compare Theorem 8 of [Br].

PROPOSITION 7.3.E. *If (7.3.24) is a strongly parabolic scalar equation, the solution guaranteed by Proposition 7.3.C exists for all $t > 0$.*

PROOF: An L^∞-bound on $u(t)$ follows from the maximum principle, and then (7.3.25) gives a C^r-bound on $u(t)$, for some $r > 0$. Hence global existence follows from Proposition 7.3.C.

Let us also consider the parabolic analogue of the PDE (2.2.62), i.e.,

$$\partial u/\partial t = \sum A^{jk}(\nabla u)\partial_j\partial_k u, \quad u(0) = f, \tag{7.3.26}$$

with

$$A^{jk}(p) = F_{p_j p_k}(p).$$

Again assume u is scalar. Then Proposition 7.3.B applies, given $f \in H^s(M)$, $s > \frac{1}{2}n + 1$. Furthermore, $u_\ell = \partial_\ell u$ satisfies

$$\partial u_\ell/\partial t = \sum \partial_j A^{jk}(\nabla u)\partial_k u_\ell, \quad u_\ell(0) = f_\ell = \partial_\ell f. \tag{7.3.27}$$

The maximum principle applies to both (7.3.26) and (7.3.27). Thus, given $u \in C([0,T], H^s) \cap C^\infty((0,T) \times M)$,

$$|u(t,x)| \leq \|f\|_{L^\infty}, \quad |u_\ell(t,x)| \leq \|f_\ell\|_{L^\infty}, \quad 0 \leq t < T. \tag{7.3.27}$$

Now the De Georgi-Nash-Moser theory applies to (7.3.27) to yield

$$\|u_\ell(t,\cdot)\|_{C^r(M)} \leq K, \quad 0 \leq t < T, \tag{7.3.28}$$

for some $r > 0$, as long as the ellipticity hypothesis from (2.2.61) applies. Hence again there is global solvability:

PROPOSITION 7.3.F. *If $F(p)$ satisfies (2.2.61), then (7.3.26) has a solution for all $t > 0$, given $f \in H^s(M)$, $s > \frac{1}{2}n + 1$.*

§7.4. Semilinear parabolic systems

Here we study equations of the form

$$\partial u/\partial t = \Delta u + F(x, D_x^1 u), \quad u(0) = f. \tag{7.4.1}$$

We suppose $u(t,x)$ takes values in $\mathbb{R}^k$, $t \in [0,T)$, $x \in M$, and Δ is the Laplace operator on M, acting componentwise in u, though clearly Δ can be replaced by more general second order strongly elliptic operators. We begin by stating the following result, which is merely a specialization of Proposition 7.3.C.

PROPOSITION 7.4.A. *If $f \in H^s(M)$, $s > \frac{1}{2}n$ (and $s > 1$ if $n = 1$), then there is a unique solution to (7.4.1), $u \in C([0,T), H^s(M)) \cap C^\infty((0,T)\times M)$, which persists as long as $\|u(t)\|_{C^r}$ is bounded, given $r > 0$.*

Though this has been established, we record a proof of the persistence statement here, since in the semilinear case it simplifies. Note that

$$F(x, D_x^1 u) = A(u; x, D)u + R(u) \tag{7.4.2}$$

with $R(u) \in C^\infty$ and, for $0 < r < 1$,

$$u \in C^r \Longrightarrow A(u; x, D) \in OPS^{2-r}_{1,1}. \tag{7.4.3}$$

Now we have

$$(d/dt)\|\Lambda^s u\|^2_{L^2} = -2\|\Lambda^{s+1}u\|^2_{L^2} + 2(\Lambda^s F(x, D_x^1 u), \Lambda^s u). \tag{7.4.4}$$

We can dominate the last term by

$$\begin{aligned} 2\|\Lambda^{s+1}u\|_{L^2}\cdot\|\Lambda^{s-1}F(x, D^1u)\|_{L^2} \\ \le \|\Lambda^{s+1}u\|^2_{L^2} + \|\Lambda^{s-1}F(x, D^1u)\|^2_{L^2}. \end{aligned} \tag{7.4.5}$$

The first term on the right side of (7.4.5) can be absorbed into the first term on the right side of (7.4.4). Meanwhile, (7.4.2)-(7.4.3) yield

$$\begin{aligned} \|\Lambda^{s-1}F(x, D^1u)\|^2_{L^2} &\le C(\|u\|_{C^r})\big[\|u\|^2_{H^{s+1-r}} + 1\big] \\ &\le \frac{1}{2}\|u\|^2_{H^{s+1}} + C'(\|u\|_{C^r})\big[\|u\|^2_{H^s} + 1\big] \end{aligned} \tag{7.4.6}$$

so

$$(d/dt)\|\Lambda^s u\|^2_{L^2} \le -\frac{1}{2}\|u\|^2_{H^{s+1}} + C'(\|u\|_{C^r})\big[\|\Lambda^s u\|^2_{L^2} + 1\big], \tag{7.4.7}$$

which implies the persistence given a bound on $\|u(t)\|_{C^r}$.

Note that, if we do not use (7.4.2)-(7.4.3) but instead appeal to the Moser estimate

$$\|F(x, D^1u)\|^2_{H^{s-1}} \le C(\|u\|_{C^1})\big[\|u\|^2_{H^s} + 1\big], \tag{7.4.8}$$

we still have the estimate (7.4.7) for $r = 1$. This result is of course also a special case of Proposition 7.1.C.

We derive further results on solvability of (7.4.1), making more specific use of the semilinear structure. We convert (7.4.1) to the integral equation

$$\begin{aligned} u(t) &= e^{t\Delta}f + \int_0^t e^{(t-s)\Delta}F(x, D^1u(s))\,ds \\ &= \Psi u(t). \end{aligned} \tag{7.4.9}$$

We want to construct a Banach space $C([0,T],X)$ such that Ψ acts as a contraction map on a certain closed subset, and hence possesses a unique fixed point. Suppose there are two Banach spaces X and Y of functions (or maybe distributions) with the following properties:

$$\begin{aligned} &e^{t\Delta} \text{ is a continuous semigroup on } X, \\ &e^{t\Delta} : Y \to X \text{ for } t > 0 \text{ and } \|e^{t\Delta}\|_{\mathcal{L}(Y,X)} \in L^1([0,1],dt), \\ &\Phi : X \to Y \text{ is locally Lipschitz, where } \Phi(u) = F(x, D^1 u). \end{aligned} \tag{7.4.10}$$

Given these hypotheses is it easy to show that, if first $\varepsilon > 0$ and then $T > 0$ are picked small enough, and

$$\mathcal{X} = \{u \in C([0,T],X) : u(0) = f, \|u(t) - f\|_X \le \varepsilon\}, \tag{7.4.11}$$

then $\Psi : \mathcal{X} \to \mathcal{X}$ and is a contraction map here.

As an example, let $X = C^r(M)$, $r \ge 1$, integer or not, and let $Y = C^{r-1}(M)$. It is easy to verify all the hypotheses in (7.4.10) in this case, as long as F is smooth in its arguments. We have the following result.

PROPOSITION 7.4.B. *Given $f \in C^r(M)$, $r \ge 1$, the equation* (7.4.1) *has a solution*

$$u \in C([0,T], C^r(M)) \cap C^\infty((0,T) \times M). \tag{7.4.12}$$

The only point left to establish is the smoothness result. But it is easy to see that $\Psi : \mathcal{X} \to C((0,T), C^{r+\rho}(M))$ for any $\rho < 1$. Hence $u \in C((0,T), C^{r+\rho})$. Replacing $t = 0$ by $t = t_o \in (0,T)$ and f by $u(t_o)$ and iterating this argument yields smoothness.

Note that the persistence result of Proposition 7.4.A is better than that given by the proof of the last result, which is persistence given a C^1-bound on $u(t)$. On the other hand, the solvability given only $f \in C^1$ is not obtainable from Proposition 7.4.A. The following consequence of Proposition 7.4.B is useful.

COROLLARY 7.4.C. *If* (7.4.1) *is solvable with a uniform bound $\|u(t)\|_{C^1} \le K_1$, for all $t \ge 0$, then there are uniform bounds*

$$\|u(t)\|_{C^\ell} \le K_\ell, \quad t \ge 1.$$

We now turn to an important example of a semilinear parabolic system, arising in the study of harmonic maps, first treated by Eells and Sampson.

Let M and N be compact Riemannian manifolds, $N \subset \mathbf{R}^k$. A harmonic map $u : M \to N$ is a critical point for the energy functional

$$E(u) = \int_M |\nabla u(x)|^2 dV(x), \tag{7.4.13}$$

amongst all such maps. Such a map, if smooth, is characterized as a solution to the semilinear equation

$$\Delta u - \Gamma(u)(\nabla u, \nabla u) = 0 \tag{7.4.14}$$

where $\Gamma(u)(\nabla u, \nabla u)$ is a certain quadratic form in ∇u, taking values in the normal space to N at $u(x)$. For this calculation, see [J] or [Str]. Denote the left side of (7.4.14) by $\tau(u)$; it can be shown that, given $u \in C^1(M, N)$, $\tau(u)$ is tangent to N at $u(x)$. Eells and Sampson proved the following result.

THEOREM 7.4.D. *Suppose N has negative sectional curvature everywhere. Then, given $v \in C^\infty(M, N)$, there exists a harmonic map $w \in C^\infty(M, N)$ which is homotopic to v.*

In [ES], the existence of w is established via solving the PDE

$$\partial u/\partial t = \Delta u - \Gamma(u)(\nabla u, \nabla u), \quad u(0) = v. \tag{7.4.15}$$

It is shown that, under the hypothesis of negative sectional curvature on N, there is a smooth solution to (7.4.15) for all $t \geq 0$, and that, for a sequence $t_k \to \infty$, $u(t_k)$ tends to the desired w.

Given that $\tau(u)$ is tangent to N for $u \in C^\infty(M, N)$, it follows that $u(t) : M \to N$ for each t in the interval $[0, T)$ on which the solution to (7.4.15) exists. In order to estimate $\nabla_x u$, Eells and Sampson produced a differential inequality for the energy density $e(t, x) = |\nabla_x u(t, x)|^2$. In fact, there is the identity

$$\begin{aligned} \partial e/\partial t - \Delta e = -|\nabla du|^2 &- \frac{1}{2}\langle du \cdot Ric^M(e_j), du \cdot e_j\rangle \\ &+ \frac{1}{2}\langle R^N(du \cdot e_j, du \cdot e_k) du \cdot e_k, du \cdot e_j\rangle, \end{aligned} \tag{7.4.16}$$

where $\{e_j\}$ is an orthonormal frame at $T_x M$ and we sum over repeated indices. Given that N has negative sectional curvature, this implies the inequality

$$\partial e/\partial t - \Delta e \leq ce. \tag{7.4.17}$$

If $f(t,x) = e^{-ct}e(t,x)$, we have $\partial f/\partial t - \Delta f \le 0$, and the maximum principle yields $f(t,x) \le \|f(0,\cdot)\|_{L^\infty}$, hence

$$e(t,x) \le e^{ct}\|\nabla v\|_{L^\infty}^2. \tag{7.4.18}$$

This C^1 estimate implies the global existence of a solution to (7.4.15), by Proposition 7.4.A, with $r = 1$, or by Proposition 7.4.B.

For the rest of Theorem 7.4.D, we need further bounds on u, including an improvement of (7.4.18). For the total energy

$$E(t) = \frac{1}{2}\int_M e(t,x)dV(x) = \frac{1}{2}\int_M |\nabla u|^2 dV(x) \tag{7.4.19}$$

we claim there is the identity

$$E'(t) = -\int_M |\partial u/\partial t|^2 dV(x). \tag{7.4.20}$$

Indeed, one easily obtains $E'(t) = -\int\langle u_t, \Delta u\rangle dV(x)$. Then replace Δu by $u_t + \Gamma(u)(\nabla u, \nabla u)$. Since u_t is tangent to N and $\Gamma(u)(\nabla u, \nabla u)$ is normal to N, (7.4.20) follows. The desired improvement of (7.4.18) wil be a consequence of the following estimate.

LEMMA 7.4.E. *Let $e(t,x) \ge 0$ satisfy the differential inequality* (7.4.17). *Assume that $E(t) = \frac{1}{2}\int e(t,x)dV(x)$ is bounded. Then there is a uniform estimate*

$$e(t,x) \le e^c\, K\|e(0,\cdot)\|_{L^\infty}, \quad t \ge 0, \tag{7.4.21}$$

where K depends only on the geometry of M.

PROOF: Writing $\partial e/\partial t - \Delta e = ce - g$, $g(t,x) \ge 0$, we have, for $0 \le s \le 1$,

$$\begin{aligned} e(t+s,x) &= e^{s(\Delta+c)}e(t,x) - \int_0^s e^{(s-\tau)(\Delta+c)}g(\tau,x)d\tau \\ &\le e^{s(\Delta+c)}e(t,x). \end{aligned} \tag{7.4.22}$$

Since $e^{s(\Delta+c)}$ is uniformly bounded from $L^1(M)$ to $L^\infty(M)$ for $s \in [\frac{1}{2}, 1]$, the bound (7.4.21) for $t \in [\frac{1}{2}, \infty)$ follows from the hypothesized L^1-bound on $e(t)$.

We remark that a more elaborate argument, which can be found on pp. 84-86 of [J], yields a explicit bound K depending on the injectivity radius of M and the first (nonzero) eigenvalue of the Laplace operator on M.

Since Lemma 7.4.E applies to $e(t,x) = |\nabla u|^2$ when u solves (7.4.15), we see that solutions to (7.4.15) satisfy

$$\|u(t)\|_{C^1} \le K_1 \|v\|_{C^1}, \text{ for all } t \ge 0. \tag{7.4.23}$$

Hence, by Corollary 7.4.C, there are uniform bounds

$$\|u(t)\|_{C^\ell} \le K_\ell \|v\|_{C^1}, \quad t \ge 1, \tag{7.4.24}$$

for each $\ell < \infty$. Of course there are consequently also uniform Sobolev bounds.

Now, by (7.4.20), $E(t)$ is positive and monotone decreasing as $t \nearrow \infty$. Thus $\int_M |u_t(t,x)|^2 dV(x)$ is an integrable function of t, so there exists a sequence $t_j \to \infty$ such that

$$\|u_t(t_j,\cdot)\|_{L^2} \to 0. \tag{7.4.25}$$

From (7.4.24) and the PDE (7.4.15) we have bounds

$$\|u_t(t,\cdot)\|_{H^k} \le C_k$$

and interpolation with (7.4.25) then gives

$$\|u_t(t_j,\cdot)\|_{H^\ell} \to 0. \tag{7.4.26}$$

Therefore, by the PDE, one has for $u_j(x) = u(t_j,x)$,

$$\Delta u_j - \Gamma(u_j)(\nabla u_j, \nabla u_j) \to 0 \text{ in } H^{\ell-2}(M), \tag{7.4.27}$$

as well as a uniform bound from (7.4.24). It easily follows that a subsequence converges in a strong norm to an element $w \in C^\infty(M,N)$ solving (7.4.14) and homotopic to v, which completes the proof of Theorem 7.4.D.

It is fairly easy to go on to show that there is an energy minimizing harmonic map $w : M \to N$ within each homotopy class, when N has negative sectional curvature, but Hartman has established a much stronger result, on the essential uniqueness of harmonic maps; cf. [J], §3.4.

We pursue a little more the method used to establish Proposition 7.4.B, in the case when $F(x, D_x^1 u)$ has extra structure such as is possessed by $\Gamma(u)(\nabla u, \nabla u)$ appearing in (7.4.15). Thus we assume

$$F(x, D_x^1 u) = B(u)(\nabla u, \nabla u), \tag{7.4.28}$$

a quadratic form in ∇u. In this case, we take

$$X = H^{1,p}, \quad Y = L^q, \quad q = p/2, \quad p > n, \tag{7.4.29}$$

and verify the three parts of the hypothesis (7.4.10), using the Sobolev imbedding result

$$H^{s,p} \subset L^{np/(n-sp)}, \ p < n/s.$$

This yields the following.

PROPOSITION 7.4.F. *If (7.4.28) is a quadratic form in ∇u, then the PDE*

$$\partial u/\partial t = \Delta u + B(u)(\nabla u, \nabla u), \quad u(0) = f, \tag{7.4.30}$$

has a solution

$$u \in C([0,T], H^{1,p}) \cap C^\infty((0,T) \times M),$$

provided

$$f \in H^{1,p}(M), \ p > n.$$

The smoothness is established by the same sort of arguments as described before. Of course the proof of Proposition 7.4.F yields persistence of solutions as long as $\|u(t)\|_{H^{1,p}}$ is bounded for some $p > n$, but again this is contained in the persistence result of Proposition 7.4.A.

Chapter 8: Nonlinear elliptic boundary problems

We establish estimates and regularity for solutions to nonlinear elliptic boundary problems. In §8.1 we treat completely nonlinear second order equations, obtaining L^2-Sobolev estimates for solutions assumed a priori to belong to $C^{2+r}(\overline{M})$, $r > 0$. The analysis here is done on a quite general level, and extends readily to higher order elliptic systems, by amalgamating the nonlinear analysis in §8.1 with the approach to linear elliptic boundary problems taken in Chapter 5 of [T2]. In §8.2 we make note of improved estimates for solutions to quasilinear second order equations. In §8.3 we show how such results, when supplemented by the DeGiorgi-Nash-Moser theory, apply to solvability of the Dirichlet problem for certain quasilinear elliptic PDE.

§8.1. Second order elliptic equations

We examine regularity near the boundary for solutions to a completely nonlinear second order elliptic PDE, with boundary condition to be prescribed later. Having looked at interior regularity in §2.2 and §3.3, we restrict attention to a collar neighborhood of the boundary $\partial M = X$, so we look at a PDE of the form

$$\partial_y^2 u = F(y, x, D_x^2 u, D_x^1 \partial_y u), \tag{8.1.1}$$

with $y \in [0,1], x \in X$. We set

$$v_1 = \Lambda u, \quad v_2 = \partial_y u, \tag{8.1.2}$$

and produce a first order system for $v = (v_1, v_2)$,

$$\begin{aligned} \partial v_1/\partial y &= \Lambda v_2 \\ \partial v_2/\partial y &= F(y, x, D_x^2 \Lambda^{-1} v_1, D_x^1 v_2). \end{aligned} \tag{8.1.3}$$

Now if we treat y as a parameter and apply the construction of §3.3 to the family of operators on functions of x, we obtain

(8.1.4)

$$F(y, x, D_x^2 \Lambda^{-1} v_1, D_x^1 v_2) = A_1(v; y, x, D_x) v_1 + A_2(v; y, x, D_x) v_2 + R(v)$$

with (for fixed y) $R(v) \in C^\infty(X)$,

$$A_j(v; y, x, \xi) \in \mathcal{A}_0^r S_{1,1}^1 \subset C^r S_{1,0}^1 \cap S_{1,1}^1 \tag{8.1.5}$$

and

$$D_x^\beta A_j \in S_{1,1}^1 \text{ for } |\beta| \le r, \quad S_{1,1}^{1+(|\beta|-r)} \text{ for } |\beta| > r, \tag{8.1.6}$$

provided $v(y,\cdot) \in C^{r+1}$. $D_y^j A_k$ has similar properties if $D_y^j v(y,\cdot)$ belongs to C^{r+1}. Doing this would lead to less than optimal results, so what we do instead is modify the construction of A_k, smoothing in both x and y, to obtain (8.1.4)-(8.1.5), with the understanding that the symbol classes reflect estimates on $D_{y,x}$ derivatives, and (8.1.6) is improved to

$$D_y^j D_x^\beta A_k \in \begin{cases} S^1_{1,1} & \text{for } j+|\beta| \le r, \\ S^{1+(j+|\beta|-r)}_{1,1} & \text{for } j+|\beta| > r, \end{cases} \tag{8.1.7}$$

if $v \in C^{1+r+}$.

Here, C^{1+r+} denotes $C^{1+r+\varepsilon}$ for some small fixed $\varepsilon > 0$. The reason for bringing this in is that an operator like $T = D_x^2 \Lambda^{-1}$ does not map $C^{1+r}(I \times X)$ to $C^r(I \times X)$, but we do have

$$T : C^{1+r+}(I \times X) \longrightarrow C^r(I \times X). \tag{8.1.7A}$$

Thus we will take $v \in C^{1+r+}$ below; we will be able to drop the '+' in the statements of the main results.

Note that, if we write $F = F(y,x,\zeta,\eta)$, $\zeta_\alpha = D_x^\alpha u$ $(|\alpha| \le 2)$, $\eta_\alpha = D_x^\alpha \partial_y u$ $(|\alpha| \le 1)$, then we can set

$$B_1(v;y,x,\xi) = \sum_{|\alpha|\le 2} (\partial F/\partial \zeta_\alpha)(D_x^2 \Lambda^{-1} v_1, D_x^1 v_2) \xi^\alpha \langle \xi \rangle^{-1} \tag{8.1.8}$$

(suppressing the y, x arguments of F) and

$$B_2(v;y,x,\xi) = \sum_{|\alpha|\le 1} (\partial F/\partial \eta_\alpha)(D_x^2 \Lambda^{-1} v_1, D_x^1 v_2) \xi^\alpha . \tag{8.1.9}$$

Thus

$$v \in C^{1+r+} \Longrightarrow A_j - B_j \in C^r S^{1-r}_{1,1}. \tag{8.1.10}$$

Using (8.1.4), we can rewrite the system (8.1.3) as

$$\begin{aligned} \partial v_1/\partial y &= \Lambda v_2 \\ \partial v_2/\partial y &= A_1(x,D)v_1 + A_2(x,D)v_2 + R(v) \end{aligned} \tag{8.1.11}$$

We also write this as

$$\partial v/\partial y = K(v;y,x,D_x)v + R \quad (R \in C^\infty) \tag{8.1.12}$$

where $K(v; y, x, D_x)$ is a 2×2 matrix of first order pseudodifferential operators. Let us denote the symbol obtained by replacing A_j by B_j as $\tilde{K}$, so

$$K - \tilde{K} \in C^r S^{1-r}_{1,1}.$$

The ellipticity condition can be expressed as

$$\text{spec } \tilde{K}(v; y, x, \xi) \subset \{z \in \mathbb{C} : |\text{Re } z| \geq C|\xi|\}, \tag{8.1.13}$$

for $|\xi|$ large. Hence we can make the same statement about the spectrum of the symbol K, for $|\xi|$ large, provided $v \in C^{1+r+}$ with $r > 0$.

In order to derive L^2-Sobolev estimates, we will construct a symmetrizer, in a fashion similar to §7.2. In particular we will make use of Lemma 7.2.B. Let $\tilde{E} = \tilde{E}(v; y, x, \xi)$ denote the projection onto the $\{\text{Re } z > 0\}$ spectral space of $\tilde{K}$, defined by

$$\tilde{E}(y, x, \xi) = \frac{1}{2\pi i} \int_\gamma \left(z - \tilde{K}(y, x, \xi)\right)^{-1} dz, \tag{8.1.14}$$

where γ is a curve enclosing that part of the spectrum of $\tilde{K}(y, x, \xi)$ contained in $\{\text{Re } z > 0\}$. Then the symbol

$$\tilde{A} = (2\tilde{E} - 1)\tilde{K} \in C^r S^1_{cl} \tag{8.1.15}$$

has spectrum in $\{\text{Re } z > 0\}$. Let $\tilde{P} \in C^r S^0_{cl}$ be a symmetrizer for the symbol $\tilde{A}$, i.e., $\tilde{P}$ and $(\tilde{P}\tilde{A} + \tilde{A}^*\tilde{P})$ are positive definite symbols, for $|\xi| \geq 1$. Then obtain (with $0 < \delta < 1$)

$$P(y, x, \xi) \in S^0_{1,\delta}; \quad P - \tilde{P} \in C^r S^{-r\delta}_{1,\delta} \tag{8.1.16}$$

by smoothing $\tilde{P}$, in (y, x). We set

$$Q = \frac{1}{2}\left(P(y, x, D_x) + P(y, x, D_x)^*\right) + K\Lambda^{-1} \tag{8.1.17}$$

with $K > 0$ picked to make the operator Q positive definite on $L^2(X)$. Similarly, define A and E by smoothing $\tilde{A}$ and $\tilde{E}$ in (y, x), so

$$\begin{aligned} &A(y, x, \xi) \in S^1_{1,\delta}, \quad A - \tilde{A} \in C^r S^{1-r\delta}_{1,\delta}, \\ &E(y, x, \xi) \in S^0_{1,\delta}, \quad E - \tilde{E} \in C^r S^{-r\delta}_{1,\delta}, \end{aligned} \tag{8.1.18}$$

and we smooth K, writing

$$K = K_0 + K^b; \quad K_0 \in S^1_{1,\delta}, \ K^b \in C^r S^{1-r\delta}_{1,\delta} \cap S^{1-r\delta}_{1,1}. \tag{8.1.19}$$

Consequently, on the symbol level,

$$
\begin{aligned}
A &= (2E-1)K_0 + A^b, \quad A^b \in S^{1-r\delta}_{1,\delta} \\
PA + A^*P &\geq C|\xi| \text{ for } |\xi| \text{ large.}
\end{aligned} \tag{8.1.20}
$$

Let us note that the homogeneous symbols $\tilde{K}$, $\tilde{E}$, and $\tilde{A}$ commute, for each (y, x, ξ); hence the commutators of the various symbols K, E, A have order $\leq r\delta$ units less than the sum of the orders of these symbols, e.g.,

$$
[E(y,x,\xi), K_0(y,x,\xi)] \in S^{1-r\delta}_{1,\delta}. \tag{8.1.21}
$$

Using this symmetrizer construction, we will look for estimates for solutions to a system of the form (8.1.12) in the spaces $H_{k,s}(M) = H_{k,s}(I \times X)$, with norms

$$
\|v\|^2_{k,s} = \sum_{j=0}^{k} \|\partial_y^j \Lambda^{k-j+s} v(y)\|^2_{L^2(I\times X)}. \tag{8.1.22}
$$

We shall differentiate $(Q\Lambda^s Ev, \Lambda^s Ev)$ and $(Q\Lambda^s(1-E)v, \Lambda^s(1-E)v)$ with respect to y (these expressions being $L^2(X)$-inner products) and sum the two resulting expressions, to obtain the desired a priori estimates, parallel to the treatment in §5.2 of [T2], following [Po].

Using (8.1.12), we have

$$
\begin{aligned}
\frac{d}{dy}(Q\Lambda^s Ev, \Lambda^s Ev) &= 2 \operatorname{Re} (Q\Lambda^s E(Kv+R), \Lambda^s Ev) \\
&\quad + (Q'\Lambda^s Ev, \Lambda^s Ev) + 2 \operatorname{Re} (Q\Lambda^s E'v, \Lambda^s Ev).
\end{aligned} \tag{8.1.23}
$$

Note that, given $v \in C^{1+r+}, r > 0$, Q' and E' belong to $OPS^{\delta}_{1,\delta}$; in fact, as in (8.1.7), we could say they belong to $OPS^{\gamma}_{1,\delta}$ with $\gamma = \max(0, (1-r)\delta)$, but this slight improvement will not be important for us here. Hence, for fixed y, each of the last two terms is bounded by

$$
C\|v(y)\|^2_{H^{s+\frac{1}{2}\delta}}. \tag{8.1.24}
$$

Here and below, we will adopt the convention that $C = C(\|v\|_{C^{1+r+}})$. Next,

$$
\begin{aligned}
(Q\Lambda^s E(Kv+R), \Lambda^s Ev) &= (Q\Lambda^s EK_0 v, \Lambda^s Ev) \\
&\quad + (Q\Lambda^s K^b v, \Lambda^s Ev) + (Q\Lambda^s ER, \Lambda^s Ev),
\end{aligned} \tag{8.1.25}
$$

where the last term is harmless and, for fixed y,

$$|(Q\Lambda^s EK^b v, \Lambda^s Ev)| \le C\|v(y)\|^2_{H^{s+\frac{1}{2}(1-r\delta)}}, \tag{8.1.26}$$

provided $s + \frac{1}{2}(1-r\delta) - (1-r\delta) > -(1-\delta)r$, i.e.,

$$s > \frac{1}{2} - r + \frac{1}{2}r\delta, \tag{8.1.27}$$

in view of (8.1.19).

Now, since $\tilde{E}(y,x,\xi)$ is a projection, we have $E(y,x,\xi)^2 - E(y,x,\xi) \in S^{-r\delta}_{1,\delta}$, and

$$\begin{aligned} E(y,x,D) - E(y,x,D)^2 = F(y,x,D) \in OPS^{-\sigma}_{1,\delta}, \\ \sigma = \min(r\delta, 1-\delta). \end{aligned} \tag{8.1.28}$$

Thus

$$QEK_0 = QAE + G; \quad G(y) \in OPS^{1-\sigma}_{1,\delta}. \tag{8.1.29}$$

Consequently we can write the first term on the right side of (8.1.25) as

$$(QAE\Lambda^s v, \Lambda^s Ev) - (G\Lambda^s v, \Lambda^s Ev) + (Q[\Lambda^s, EK_0]v, \Lambda^s Ev). \tag{8.1.30}$$

The last two terms in (8.1.30) are bounded (for each y) by

$$C\|v(y)\|^2_{H^{s+\frac{1}{2}(1-\sigma)}}. \tag{8.1.31}$$

As for the contribution of the first term in (8.1.30) to the estimation of (8.1.23), we have, for each y,

$$(QAE\Lambda^s v, \Lambda^s Ev) = (QA\Lambda^s Ev, \Lambda^s Ev) + (QA[E,\Lambda^s]v, \Lambda^s v), \tag{8.1.32}$$

the last term estimable by (8.1.31), and

$$2\,\mathrm{Re}\,(QA\Lambda^s Ev, \Lambda^s Ev) \ge C_1\|Ev(y)\|^2_{H^{s+\frac{1}{2}}} - C_2\|Ev(y)\|^2_{H^s}, \tag{8.1.33}$$

by (8.1.20) and Gårding's inequality. Keeping track of the various ingredients in the analysis of (8.1.23), we see that

$$\begin{aligned} &\frac{d}{dy}(Q\Lambda^s Ev, \Lambda^s Ev) \\ &\qquad \ge C_1\|Ev(y)\|^2_{H^{s+\frac{1}{2}}} - C_2\|v(y)\|^2_{H^{s+\frac{1}{2}(1-\sigma)}} - C_3\|R(y)\|^2_{H^s}, \end{aligned} \tag{8.1.34}$$

where $C_j = C_j(\|v\|_{C^{1+r+}}) > 0$.

A similar analysis gives

$$\begin{aligned}(8.1.35)\quad &\frac{d}{dy}(Q\Lambda^s(1-E)v, \Lambda^s(1-E)v)\\ &\quad \le -C_1\|(1-E)v(y)\|^2_{H^{s+\frac{1}{2}}} + C_2\|v(y)\|^2_{H^{s+\frac{1}{2}(1-\sigma)}} + C_3\|R(y)\|^2_{H^s}.\end{aligned}$$

Putting together these two estimates yields

$$\begin{aligned}(8.1.36)\quad &\frac{1}{2}C_1\|v(y)\|^2_{H^{s+\frac{1}{2}}} \le C_1\|Ev(y)\|^2_{H^{s+\frac{1}{2}}} + C_1\|(1-E)v(y)\|^2_{H^{s+\frac{1}{2}}}\\ &\qquad \le \frac{d}{dy}(Q\Lambda^s Ev, \Lambda^s Ev) - \frac{d}{dy}(Q\Lambda^s(1-E)v, \Lambda^s(1-E)v)\\ &\qquad\qquad + C_2\|v(y)\|^2_{H^{s+\frac{1}{2}(1-\sigma)}} + C_3\|R(y)\|^2_{H^s}.\end{aligned}$$

Now standard arguments allow us to replace $H^{s+\frac{1}{2}(1-\sigma)}$ by H^t, with $t << s$. Then integration over $y \in [0,1]$ gives

$$\begin{aligned}(8.1.37)\quad &C_1\|v\|^2_{0,s+\frac{1}{2}} \le \|\Lambda^s Ev(1)\|^2_{L^2} + \|\Lambda^s(1-E)v(0)\|^2_{L^2} + C_2\|v\|^2_{0,t} + C_3\|R\|^2_{0,s}.\end{aligned}$$

Recalling that

$$(8.1.38)\qquad \|v\|^2_{1,s} = \|\Lambda^{1+s}v\|^2_{L^2(M)} + \|\Lambda^s\partial_y v\|^2_{L^2(M)}$$

and using (8.1.12) to estimate $\partial_y v$, we have

$$(8.1.39)\quad \|v\|^2_{1,s-\frac{1}{2}} \le C\big[\|Ev(1)\|^2_{H^s} + \|(1-E)v(0)\|^2_{H^s} + \|v\|^2_{0,t} + \|R\|^2_{0,s}\big],$$

with $C = C(\|v\|_{C^{1+r+}})$, provided $v \in C^{1+r+}$ with $r > 0$ and that s satisfies the lower bound (8.1.27). Let us note that $C_1[\|\Lambda^s(1-E)v(1)\|^2_{L^2} + \|\Lambda^s Ev(0)\|^2_{L^2}]$ could have been included on the left side of (8.1.37), so we also have the estimate

$$(8.1.40)\qquad \|(1-E)v(1)\|^2_{H^s} + \|Ev(0)\|^2_{H^s} \le \text{ right side of (8.1.39)}.$$

Having completed a first round of a priori estimates, we bring in a consideration of boundary conditions that might be imposed. Of course, the boundary conditions $Ev(1) = f_1, (1-E)v(0) = f_0$ are a possibility, but these are really a tool with which to analyze other, more naturally occurring boundary conditions. The 'real' boundary conditions of interest include the Dirichlet condition on (8.1.1):

$$u(0) = f_0, \quad u(1) = f_1, \tag{8.1.41}$$

various sorts of (possibly nonlinear) conditions involving first order derivatives:

$$G_j(x, D^1 u) = f_j \text{ at } y = j \ (j = 0, 1) \tag{8.1.42}$$

and, when (8.1.1) is itself a $K \times K$ system, other possibilities, which can be analyzed in the same spirit. Now if we write $D^1 u = (u, \partial_x u, \partial_y u) = (\Lambda^{-1} v_1, \partial_x \Lambda^{-1} v_1, v_2)$, and use the construction of §3.3, we can write (8.1.42) as

$$H_j(v; x, D)v = g_j \text{ at } y = j, \tag{8.1.43}$$

where, given $v \in C^{1+r+}$,

$$H_j(v; x, \xi) \in \mathcal{A}_0^{1+r} S^0_{1,1} \subset C^{1+r} S^0_{1,0} \cap S^0_{1,1}. \tag{8.1.44}$$

Of course, (8.1.41) can be written in the same form, with $H_j v = v_1$.

Now the following is the natural regularity hypothesis to make on (8.1.43), namely that we have an estimate of the form

(8.1.45)
$$\sum_j \|v(j)\|^2_{H^s} \le C\Big[\|Ev(0)\|^2_{H^s} + \|(1-E)v(1)\|^2_{H^s}\Big] + C\sum_j \Big[\|H_j(v; x, D)v(j)\|^2_{H^s} + \|v(j)\|^2_{H^{s-1}}\Big].$$

We then say the boundary condition is *regular*. If we combine this with (8.1.39) and (8.1.40), we obtain the following fundamental estimate.

PROPOSITION 8.1.A. *If v satisfies the elliptic system (8.1.12), together with the boundary condition (8.1.43), assumed to be regular, then*

$$\|v\|^2_{1,s-\frac{1}{2}} \le C\Big[\sum_j \|g_j\|^2_{H^s} + \|v\|^2_{0,t} + \|R\|^2_{0,s}\Big], \tag{8.1.46}$$

provided $v \in H_{1,s-\frac{1}{2}} \cap C^{1+r}, r > 0$, and s satisfies (8.1.27). We can take $t << s$. In case (8.1.42) holds, we can replace $\|g_j\|_{H^s}$ by $\|f_j\|_{H^s}$, and in *case the Dirichlet condition* (8.1.41) *holds and is regular, we can replace* $||g_j||_{H^s}$ *by* $||f_j||_{H^{s+1}}$, *in* (8.1.46).

We can now easily obtain higher order estimates, of the form

$$\|v\|^2_{k,s-\frac{1}{2}} \leq C\Big[\sum_j \|g_j\|^2_{H^{s+k-1}} + \|v\|^2_{0,t} + \|R\|^2_{k-1,s}\Big] \tag{8.1.47}$$

for $t << s - \frac{1}{2}$, by induction from

$$\|v\|^2_{k,s-\frac{1}{2}} = \|v\|^2_{k-1,s+\frac{1}{2}} + \|\partial_y v\|^2_{k-1,s-\frac{1}{2}}, \tag{8.1.48}$$

plus substituting the right side of (8.1.12) for ∂_y. We also obtain a corresponding regularity theorem, via inclusion of Friedrichs mollifiers in the standard fashion. Thus replace Λ^s by $\Lambda^s_\varepsilon = \Lambda^s J_\varepsilon$ in (8.1.23) and churn away. One must keep in mind that K^b must be applicable to $v(y)$ for the analogue of (8.1.26) to work. Given (8.1.19), we need $v(y) \in H^\sigma$ with $\sigma > 1 - r$. However, $v \in C^{1+r}$ already implies this. We thus have the following result.

THEOREM 8.1.B. *Let* v *be a solution to the elliptic system* (8.1.12), *satisfying boundary conditions* (8.1.43), *assumed to be regular. Assume*

$$v \in C^{1+r}, \ r > 0, \tag{8.1.49}$$

$$g_j \in H^{s+k-1}(X), \ s > \frac{1}{2} - r + \frac{1}{2} r\delta. \tag{8.1.50}$$

Then

$$v \in H_{k,s-\frac{1}{2}}(I \times X). \tag{8.1.51}$$

In particular, taking $s = \frac{1}{2}$, and noting that

$$H_{k,0}(M) = H^k(M), \tag{8.1.52}$$

we can specialize (8.1.50)-(8.1.51) to

$$g_j \in H^{k-\frac{1}{2}}(X) \Longrightarrow v \in H^k(I \times X), \tag{8.1.53}$$

for $k = 1, 2, 3, \dots$, granted (8.1.49) (which makes the $k = 1$ case trivial).

Note that, in (8.1.34)-(8.1.36), one could replace the term $||R(y)||^2_{H^s}$ by $||R(y)||_{H^{s-\frac{1}{2}}} \cdot ||v(y)||_{H^{s+\frac{1}{2}}}$; then an absorption can be performed in

(8.1.36), and hence in (8.1.37)-(8.1.39) we can substitute $\|R\|^2_{0,s-\frac{1}{2}}$, and use $\|R\|^2_{k-1,s-\frac{1}{2}}$ in (8.1.47).

We note that Theorem 8.1.B is also valid for solutions to a non-homogeneous elliptic system, where R in (8.1.12) can contain an extra term, belonging to $H_{k-1,s-\frac{1}{2}}$, and then the estimate (8.1.47), strengthened as indicated above, and consequent regularity theorem are still valid. If (8.1.1) is generalized to

$$\partial_y^2 u = F(D_x^2 u, D_x^1 \partial_y u) + f, \tag{8.1.54}$$

then a term of the form $(0, f)^t$ is added to (8.1.12).

In view of the estimate (8.1.10) comparing the symbol of K with that obtained from the linearization of the original PDE (8.1.1), and the analogous result that holds for H_j, derived from G_j, we deduce the following.

PROPOSITION 8.1.C. *Suppose that, at each point on ∂M, the linearization of the boundary condition of* (8.1.42) *is regular for the linearization of the PDE* (8.1.1). *Then the regularity estimate* (8.1.42) *holds. In particular, this holds for the Dirichlet problem, for any scalar (real) elliptic PDE of the form* (8.1.1).

It is useful to make the following technical remark on the analysis done above. Namely, we can replace the hypothesis $v \in C^{1+r+}$ by $u \in C^{2+r}$ in (8.1.7), (8.1.10), and subsequent places, thus eliminating reference to C^{1+r+}.

§8.2. Quasilinear elliptic equations

We establish here a strengthened version of Theorem 8.1.B when u solves a quasilinear second order elliptic PDE, with a regular boundary condition. Thus we are looking at the special case of (8.1.1) in which

(8.2.1)
$$F(y, x, D_x^2 u, D_x^1 \partial_y u) = -\sum_j B^j(x, y, D^1 u)\partial_j \partial_y u \\ - \sum_{j,k} A^{jk}(x, y, D^1 u)\partial_j \partial_k u + F_1(x, y, D^1 u).$$

All the calculations of §8.1 apply, but some of the estimates are better. This is because when we derive the equation (8.1.12), i.e.,

$$\partial v/\partial y = K(v; y, x, D_x)v + R \quad (R \in C^\infty) \tag{8.2.2}$$

for $v = (v_1, v_2) = (\Lambda u, \partial_y u)$, (8.1.5) is improved to

$$u \in C^{r+1} \Longrightarrow K \in \mathcal{A}_0^r S^1_{1,1} + S^{1-r}_{1,1} \quad (r > 0). \tag{8.2.3}$$

Compare (3.3.23). Under the hypothesis $u \in C^{r+1}$, one has the result (8.1.15), $\tilde{A} \in C^r S^1_{cl}$, which before required $u \in C^{2+r}$. Also (8.1.18)-(8.1.20) now hold for $u \in C^{1+r}$. Thus all the a priori estimates, down through (8.1.47), hold, with $C = C(\|u\|_{C^{1+r}})$. One point that must be taken into consideration is that, for the estimates to work, one needs $v(y) \in H^\sigma$ with $\sigma > 1 - r$, and now this does not necessarily follow from the hypothesis $u \in C^{1+r}$. Hence we have the following regularity result. Compare the interior regularity established in Theorem 2.2.E.

THEOREM 8.2.A. *Let u satisfy a second order quasilinear elliptic PDE with a regular boundary condition, of the form* (8.1.43), *for $v = (\Lambda u, \partial_y u)$. Assume that*

$$u \in C^{1+r} \cap H_{1,\sigma}, \quad r > 0, \quad r + \sigma > 1. \tag{8.2.4}$$

Then, for $k = 0, 1, 2, \dots$,

$$g_j \in H^{k-\frac{1}{2}}(X) \Longrightarrow v \in H^k(I \times X). \tag{8.2.5}$$

The Dirichlet boundary condition is regular (if the PDE is real and scalar) and

$$u(j) = f_j \in H^{k+s}(X) \Longrightarrow v \in H_{k,s-\frac{1}{2}}(I \times X), \tag{8.2.6}$$

if $s > \frac{1}{2}(1 - r)$. In particular,

$$\begin{aligned} u(j) = f_j \in H^{k+\frac{1}{2}}(X) &\Longrightarrow v \in H^k(I \times X) \\ &\Longrightarrow u \in H^{k+1}(I \times X). \end{aligned} \tag{8.2.7}$$

We consider now the further special case

$$\begin{aligned} F(y, x, D_x^2 u, D_x^1 \partial_y u) = &- \sum_j B^j(x, y, u) \partial_j \partial_y u \\ &- \sum_{j,k} A^{jk}(x, y, u) \partial_j \partial_k u + F_1(x, y, D^1 u). \end{aligned} \tag{8.2.8}$$

In this case, when we derive the system (8.2.2), we have the implication

$$u \in C^r(\overline{M}) \Longrightarrow K \in \mathcal{A}_0^r S^1_{1,1} + S^{1-r}_{1,1} \quad (r > 0). \tag{8.2.9}$$

Similarly, under this hypothesis we have $\tilde{A} \in C^r S^1_{cl}$, etc. Therefore we have the following.

PROPOSITION 8.2.B. *If u satisfies the PDE* (8.1.1) *with F given by* (8.2.8), *then the conclusions of Theorem* 8.2.A *hold when the hypothesis* (8.2.4) *is weakened to*

$$u \in C^r \cap H_{1,\sigma}, \quad r + \sigma > 1. \tag{8.2.10}$$

Note that associated to this regularity is an estimate. For example, if u satisfies the Dirichlet boundary condition, we have, for $k \geq 2$,

$$\|u\|_{H^k(M)} \leq C_k(\|u\|_{C^r(\overline{M})})\big[\|u|_{\partial M}\|_{H^{k-\frac{1}{2}}(\partial M)} + \|u\|_{L^2(M)}\big], \tag{8.2.11}$$

where we have used Poincare's inequality to replace the $H_{1,\sigma}$-norm of u by the L^2-norm on the right.

§8.3. Interface with DeGiorgi-Nash-Moser theory

We resume the discussion begun at the end of §2.2 of a class of quasi-linear elliptic PDEs whose study involves first the DeGiorgi-Nash-Moser regularity theory and then Schauder type extimates. The version of Theorem 2.2.J for bounded regions, with Dirichlet boundary conditions, is the following. Suppose $\overline{M}$ is compact with smooth boundary ∂M.

THEOREM 8.3.A. *Let $u \in H^1(M)$ solve the scalar PDE*

$$\sum \partial_j a_{jk}(x) \partial_k u = g + \sum \partial_j f_j \text{ on } M, \tag{8.3.1}$$

with $g \in L^{q/2}$, $f_j \in L^q$, $q > n = \dim M$, and $a_{jk} \in L^\infty(M)$ satisfying

$$\lambda_0 |\xi|^2 \leq \sum a_{jk}(x) \xi_j \xi_k \leq \lambda_1 |\xi|^2, \tag{8.3.2}$$

for constants $\lambda_j > 0$. Suppose $u|_{\partial M} = \varphi \in C^1(\partial M)$. Then $u \in C^r(\overline{M})$ for some $r > 0$, and

$$\|u\|_{C^r} \leq K(\lambda_0, \lambda_1, M)\big[\|g\|_{L^{q/2}} + \sum \|f_j\|_{L^q} + \|\varphi\|_{C^1(\partial M)}\big]. \tag{8.3.3}$$

A proof of this is given in Appendix C.

We want to establish existence of smooth solutions to the nonlinear elliptic PDE

$$\Phi(D^2 u) = \sum F_{p_j p_k}(\nabla u) \partial_j \partial_k u = 0 \text{ on } M, \quad u = \varphi \text{ on } \partial M, \tag{8.3.4}$$

which we derived in (2.2.62) as a PDE satisfied by the minimizer of

$$I(u) = \int_M F(\nabla u)dx \tag{8.3.5}$$

over the space V_φ^1. Assume $\varphi \in C^\infty(\overline{M})$. We continue to assume F is smooth and satisfies

$$\begin{aligned} C_1|p|^2 - L_1 \le F(p) \le C_2|p|^2 + K_2, \\ A_1|\xi|^2 \le \sum F_{p_jp_k}(p)\xi_j\xi_k \le A_2|\xi|^2. \end{aligned} \tag{8.3.6}$$

We use the *method of continuity*, showing that, for each $\tau \in [0,1]$, there is a smooth solution to

$$\Phi_\tau(D^2u) = 0 \text{ on } M, \quad u = \varphi \text{ on } \partial M, \tag{8.3.7}$$

where

$$\begin{aligned} \Phi_\tau(D^2u) &= \tau\Phi(D^2u) + (1-\tau)\Delta u \\ &= \sum A_\tau^{jk}(\nabla u)\partial_j\partial_k u \end{aligned} \tag{8.3.8}$$

with

$$A_\tau^{jk}(\nabla u) = \tau F_{p_jp_k}(\nabla u) + (1-\tau)\delta_{jk}. \tag{8.3.9}$$

Clearly (8.3.7) is solvable for $\tau = 0$. Let J be the largest interval containing $\{0\}$ such that (8.3.7) has a solution $u = u_\tau \in C^\infty(\overline{M})$ for each $\tau \in J$. We will show that J is all of $[0,1]$ by showing it is both open and closed in $[0,1]$. The openness is the relatively easy part.

LEMMA 8.3.B. *If $\tau_0 \in J$, then, for some $\varepsilon > 0$, $[\tau_0, \tau_0 + \varepsilon) \subset J$.*

PROOF: Fix k large and define

$$\Psi : [0,1] \times V_\varphi^k \longrightarrow H^{k-2}(M) \tag{8.3.10}$$

by $\Psi(\tau, u) = \Phi_\tau(D^2u)$, where

$$V_\varphi^k = \{u \in H^k(M) : u = \varphi \text{ on } \partial M\}. \tag{8.3.11}$$

This map is C^1 and its derivative with respect to the second argument is

$$D_2\Psi(\tau_0, u)v = Lv, \tag{8.3.12}$$

where

$$L : V_0^k = H^k \cap H_0^1 \longrightarrow H^{k-2}(M) \tag{8.3.13}$$

is given by

$$Lv = \sum \partial_j A^{jk}_{\tau_0}(\nabla u(x))\partial_k v. \tag{8.3.14}$$

L is an elliptic operator with coefficients in $C^\infty(\overline{M})$, when $u = u_{\tau_0}$, clearly an isomorphism in (8.3.13). Thus, by the implicit function theorem, for τ close enough to τ_0 there will be u_τ, close to u_{τ_0}, such that $\Psi(\tau, u_\tau) = 0$. Since $u_\tau \in H^k(M)$ solves the regular elliptic boundary problem (8.3.7), if we pick k large enough we can apply the regularity result of Theorem 8.1.B to deduce $u_\tau \in C^\infty(\overline{M})$.

The next task is to show that J is closed. This will follow from a sufficient a priori bound on solutions $u = u_\tau$, $\tau \in J$. We start with fairly weak bounds. First, the maximum principle implies

$$\|u\|_{L^\infty(M)} = \|\varphi\|_{L^\infty(\partial M)}, \tag{8.3.15}$$

for each $u = u_\tau$, $\tau \in J$.

Next we estimate derivatives. For simplicity we take $\overline{M} = [0,1]\times T^{n-1}$. Each $u_\ell = \partial_\ell u$ satisfies

$$\sum \partial_j A^{jk}(\nabla u)\partial_k u_\ell = 0, \tag{8.3.16}$$

where $A^{jk}(\nabla u)$ is given by (8.3.9). The ellipticity, which follows from hypothesis (8.3.6), implies a bound

$$\|u_\ell\|_{H^1(M)} \le K, \quad 1 \le \ell \le n-1, \tag{8.3.17}$$

provided ∂_ℓ is tangent to ∂M for $1 \le \ell \le n-1$, since $u_\ell = \partial_\ell \varphi$ on ∂M. Also, a fairly elementary barrier construction bounds $\nabla u|_{\partial M}$, and then the maximum principle applied to (8.3.16) yields a uniform bound

$$\|\nabla u\|_{L^\infty(M)} \le K. \tag{8.3.18}$$

Now Theorem 8.3.A enters in the following way; it applies to $u_\ell = \partial_\ell u$, for $1 \le \ell \le n-1$. Thus there is an $r > 0$ for which we have bounds

$$\|u_\ell\|_{C^r(\overline{M})} \le K, \quad 1 \le \ell \le n-1. \tag{8.3.19}$$

Recall from the end of §2.2 that such a property on *all* first derivatives of a solution to (8.3.4) led to the applicability of Schauder estimates to establish interior regularity.

In the case of examining regularity at the boundary, more work is required, since (8.3.19) does not include a derivative ∂_n transverse to the boundary. Now, using (8.3.7), we can solve for $\partial_n^2 u$ in terms of $\partial_j \partial_k u$, for $1 \le j \le n$, $1 \le k \le n-1$. This leads, via a nontrivial argument, to the estimate

$$\|u\|_{C^{r+1}(\overline{M})} \le K. \tag{8.3.20}$$

This result, due to Morrey, will be established below.

Granted (8.3.20), we can then apply the estimate (8.2.11) to $w = \nabla u$, satisfying $\sum \partial_j A^{jk}(w) \partial_k w = 0$. Thus, for any k,

$$\|\nabla u\|_{H^k(M)} \le K_k. \tag{8.3.21}$$

Therefore, if $[0, \tau_1) \subset J$, as $\tau_\nu \nearrow \tau_1$ we can pick a subsequence of u_{τ_ν} converging weakly in $H^{k+1}(M)$, hence strongly in $H^k(M)$. If k is picked large enough, the limit u_1 is an element of $H^{k+1}(M)$, solving (8.3.7) for $\tau = \tau_1$, and furthermore the regularity result Theorem 8.2.A is applicable; hence $u_1 \in C^\infty(\overline{M})$. This shows that J is closed.

Hence, modulo a proof of Morrey's result (8.3.20), we have the proof of solvability of the boundary problem (8.3.4).

In order to prove (8.3.20), we will use the Morrey spaces, discussed in §A.2. We will show that

$$\int_{B_R \cap M} |\nabla u_\ell|^2 dx \le C R^{n-2+2r}, \quad 1 \le \ell \le n-1. \tag{8.3.22}$$

This implies (see (A.2.10)) that

$$\partial_k \partial_\ell u \in M^p(M) \text{ for } 1 \le \ell \le n-1,\ 1 \le k \le n, \tag{8.3.23}$$

where $p \in (n, \infty)$ and $r \in (0,1)$ are related by $1 - r = n/p$. Now the PDE (8.3.7) enables us to write $\partial_n^2 u$ as a linear combination of the terms in (8.3.23), with $L^\infty(M)$-coefficients. Hence $\partial_n^2 u \in M^p(M)$, so

$$\nabla(\partial_n u) \in M^p(M). \tag{8.3.24}$$

A fundamental result of Morrey is that

$$\nabla v \in M^p(M) \Longrightarrow v \in C^r(\overline{M}). \tag{8.3.25}$$

This is proved in §A.2; see Theorem A.2.A. Thus $\partial_n u \in C^r(\overline{M})$, and this together with (8.3.19) yields (8.3.20).

It remains to consider (8.3.22); we will establish such an integral estimate with u_ℓ replaced by $v_\ell = u_\ell - \varphi_\ell$, with $\varphi_\ell = \partial_\ell \varphi$. This will suffice. For v_ℓ, we have the PDE

$$\sum \partial_j A^{jk}(\nabla u)\partial_k v_\ell = \sum \partial_j f_j, \tag{8.3.26}$$

$$f_j = \sum_k A^{jk}(\nabla u)\partial_k \varphi_\ell \in L^\infty(M).$$

For any $y \in M$, a center for concentric balls B_R and B_{2R}, (which may reach outside $\overline{M}$), choose a positive function $\psi \in C_0^1(B_{2R})$ such that

$$\psi = 1 \text{ on } B_R, \quad |\nabla \psi| \le 2/R. \tag{8.3.27}$$

Pick a constant c such that

$$c = v_\ell(y) \text{ if } B_{2R} \subset M; \quad c = 0 \text{ if } B_{2R} \cap \partial M \ne \emptyset. \tag{8.3.28}$$

Hence $\psi(x)^2(v_\ell(x) - c) \in H_0^1(M)$. We estimate

$$\sum \int_M \psi(x)^2 A^{jk} \partial_j v_\ell \cdot \partial_k v_\ell dx \tag{8.3.29}$$

from above, substituting $v_\ell - c$ for v_ℓ, integrating by parts and using (8.3.26). There follows readily a bound

$$\int_M \psi(x)^2 |\nabla v_\ell|^2 dx \le C \int_M \left[\psi(x)^2 + |\nabla \psi|^2 (v_\ell - c)^2\right] dx. \tag{8.3.30}$$

Compare (2.2.67)-(2.2.68). Now the Hölder estimates (8.3.19) imply that $(v_\ell - c)^2 \le K R^{2r}$ on B_{2R}, so from (8.3.30) we get

$$\begin{aligned} \int_{B_R} |\nabla v_\ell|^2 dx &\le C[R^n + R^{n-2} \cdot R^{2r}] \\ &\le C' R^{n-2+2r}, \end{aligned} \tag{8.3.31}$$

from which (8.3.22) follows. This proves Morrey's result, (8.3.20).

As noted, to have $\partial_\ell, 1 \le \ell \le n-1$, tangent to ∂M, we require $\overline{M} = [0,1] \times T^{n-1}$. For $\overline{M} \subset \mathbb{R}^n$, if $X = \sum b_\ell \partial_\ell$ is a smooth vector field tangent to ∂M, then $u_X = Xu$ solves, in place of (8.3.16),

$$\sum \partial_j A^{jk}(\nabla u)\partial_k u_X = \sum \partial_j F_j$$

with $F_j \in L^\infty$ calculable in terms of ∇u. Thus Theorem 8.3.A still applies, and the rest of the argument above extends easily.

Chapter 9: Extension of the Schauder estimates

We describe here, from the point of view of the paradifferential calculus as used in Chapters 3 and 8, a refinement of the Schauder estimates due to Nirenberg. We also describe an important application of this to Monge-Ampere equations, given in [CNS].

§9.1. Nirenberg's refinement

In this section we will consider solutions to nonlinear elliptic equations on a compact manifold M, without boundary. The regularity result of Theorem 2.2.G for solutions to

$$F(x, D^m u) = f, \tag{9.1.1}$$

assumed to be elliptic, implies the estimate

$$\|u\|_{C_*^{m+r}} \leq C(\|u\|_{C^{m+\epsilon}})\big[\|f\|_{C_*^r} + \|u\|_{C_*^s} + 1\big] \tag{9.1.2}$$

given $\varepsilon > 0,\ s < m + r$. Similarly, given $1 < p < \infty$,

$$\|u\|_{H^{m+r,p}} \leq C(\|u\|_{C^{m+\epsilon}})\big[\|f\|_{H^{r,p}} + \|u\|_{H^{s,p}} + 1\big]. \tag{9.1.3}$$

Nirenberg [N2] showed how to replace $C(\|u\|_{C^{m+\epsilon}})$ by a factor depending on any modulus of continuity of $D^m u$, possibly somewhat weaker than any Hölder estimate on $D^m u$. We will give a proof of such an estimate here, via a modification of the analysis used to establish Theorem 3.3.C.

Given a continuous monotonic function $\varphi : [0, 1] \to \mathbb{R},\ \varphi(0) = 0$, which we will call a modulus, we will say $u \in C_\varphi$ if u is bounded and

$$|u(x) - u(y)| \leq C\varphi(|x - y|) \text{ for } |x - y| \leq 1. \tag{9.1.4}$$

We use the norm

$$\|u\|_{C_\varphi} = \|u\|_{L^\infty} + \sup_{|x-y|\leq 1} |u(x) - u(y)|/\varphi(|x - y|). \tag{9.1.5}$$

Given $m \in \mathbb{Z}^+$, we say $u \in C_\varphi^m$ if $D^\alpha u \in C_\varphi$ for $|\alpha| \leq m$. We can use the norm

$$\|u\|_{C_\varphi^m} = \sum_{|\alpha|\leq m} \|D^\alpha u\|_{C_\varphi}. \tag{9.1.6}$$

Recall the formula

$$F(x, D^m u) = M(u; x, D) + F(x, D^m \Psi_0(D)u) \tag{9.1.7}$$

with
(9.1.8)

$$M(u;x,\xi) = \sum_{|\alpha|\le m} \sum_j \int_0^1 (\partial F/\partial \zeta_\alpha)(\Psi_j D^m u + \tau \psi_{j+1} u) d\tau \cdot \psi_{j+1}(\xi)\xi^\alpha.$$

As shown in §3.3,

$$u \in C^m \Longrightarrow M(u;x,\xi) \in \mathcal{A}^0 S^m_{1,1}. \tag{9.1.9}$$

Recall the symbol smoothing of §1.3:

$$M(u;x,\xi) = M^\#(x,\xi) + M^b(x,\xi), \quad M^\#(x,\xi) \in S^m_{1,\delta}. \tag{9.1.10}$$

We choose $\delta \in (0,1)$. Thus, for $p \in (1,\infty)$, $s \in \mathbf{R}$,

$$\|M^\#(x,D)v\|_{H^{s,p}} \le C_{sp}(\|u\|_{C^m})\|v\|_{H^{s+m,p}}. \tag{9.1.11}$$

Furthermore, a parametrix $P(x,D) \in OPS^{-m}_{1,\delta}$ for $M^\#(x,D)$, as in (3.3.26), satisfies an estimate

$$\|P(x,D)v\|_{H^{s+m,p}} \le C_{sp}(\|u\|_{C^m})\|v\|_{H^{s,p}}. \tag{9.1.12}$$

Also $P(x,D)M^\#(x,D) = I + S$ where S satisfies the same sort of estimate. Now (9.1.3) results from the estimate

$$\|M^b(x,D)v\|_{H^{s,p}} \le C_{sp}(\|u\|_{C^{m+\epsilon}})\|v\|_{H^{s+m-\delta,p}} \tag{9.1.13}$$

valid for

$$0 < \delta < \varepsilon, \ s > 0.$$

It is this estimate we need to modify for $u \in C^m_\varphi$.

LEMMA 9.1.A. *If $u \in C^m_\varphi$, $1 < p < \infty$, $s > 0$, then there exist $K(\varepsilon) = K_\varphi(\varepsilon)$, such that*

$$\|M^b(x,D)v\|_{H^{s,p}} \le C_{sp}(\|u\|_{C^m_\varphi}) \cdot \big[\varepsilon\|v\|_{H^{s+m,p}} + K(\varepsilon)\|v\|_{L^p}\big]. \tag{9.1.14}$$

for $0 < \varepsilon \le 1$.

PROOF: Given $\eta \in (0,1]$, write $M^b(x,\xi) = A_\eta(x,\xi) + B_\eta(x,\xi)$, with $A_\eta(x,\xi) = M^b(x,\xi)\Phi(\eta\xi)$, given $\Phi \in C_0^\infty(\mathbf{R}^n)$, $\Phi(\xi) = 1$ for $|\xi| \le 1$, 0 for $|\xi| \ge 2$. Since $M^b(x,\xi) \in S^m_{1,1}$, we hence have a bound on $\eta^{m+s}A_\eta(x,\xi)$ in $S^{-s}_{1,1}$ for $\eta \in (0,1]$, and consequently an estimate

$$\|A_\eta(x,D)v\|_{H^{s,p}} \le C_{sp}(\|u\|_{C^m})\eta^{-m-s}\|v\|_{L^p} \tag{9.1.15}$$

for $s > 0$. For an estimate on $B_\eta(x, D)$, note that

$$\|f - J_\varepsilon f\|_{L^\infty} \le C\varphi(\varepsilon)\|f\|_{C_\varphi}. \tag{9.1.16}$$

The resulting symbol estimates on $M^b(x,\xi)$ imply

$$\|B_\eta(x, D)v\|_{H^{s,p}} \le C_{sp}(\|u\|_{C^m_\varphi})\varphi(\eta^\delta)\|v\|_{H^{s+m,p}}, \tag{9.1.17}$$

for $s > 0$. The desired estimate (9.1.14) follows upon making the change of variable $\varepsilon = \varphi(\eta^\delta)$.

We now produce a Nirenberg-type estimate for a solution to (9.1.1); suppose $f \in H^{s,p}$, $s > 0$, $1 < p < \infty$. Using $P = P(x, D)$ and S as above, we have

$$u = Pf - PM^b(x, D)u - PR(u) - Su \tag{9.1.18}$$

where $R(u)$ is the last term on the right side of (9.1.7). Take the $H^{s+m,p}$-norm of each side. From (9.1.14) we have

$$\|PM^b(x, D)u\|_{H^{s+m,p}} \le C(\|u\|_{C^m_\varphi})\big[\varepsilon\|u\|_{H^{s+m,p}} + K(\varepsilon)\|u\|_{L^p}\big]. \tag{9.1.19}$$

If we pick ε to be the smaller of 1 and $[2C(\|u\|_{C^m_\varphi})]^{-1}$, we can absorb the term involving $\|u\|_{H^{s+m,p}}$ on the right, yielding the desired estimate:

THEOREM 9.1.B. *If $u \in C^m_\varphi \cap H^{s+m,p}$ solves the elliptic PDE (9.1.1), $s > 0$, $1 < p < \infty$, then there is an a priori estimate*

$$\|u\|_{H^{s+m,p}} \le C(\|u\|_{C^m})\|f\|_{H^{s,p}} + C(\|u\|_{C^m_\varphi}) \tag{9.1.20}$$

Clearly the same reasoning yields an estimate on $\|u\|_{C^{m+s}_*}$ in terms of $\|f\|_{C^s_*}$, with similar coefficients as appear in (9.1.20).

§9.2. Elliptic boundary problems

Refinements of the sort described in the last section were noted for elliptic boundary problems in [ADN]. Here we indicate the modifications in the analysis of §8.1 needed to establish such estimates, for an elliptic second order system on $\overline{M} = I \times X$:

$$\begin{aligned} \partial_y^2 u &= F(y, x, D_x^2 u, D_x^1 \partial_y u) + f \text{ on } M, \\ G_j(x, D^1 u) &= f_j \text{ at } y = j \quad (j = 0, 1). \end{aligned} \tag{9.2.1}$$

We will assume $u \in C^2_\varphi(\overline{M})$. We use the reduction to the first order system (8.1.12) and the same construction of symmetrizer, etc., as in §8.1. The results obtained there hold, with $r = 0$, which is not very useful.

However, there are systematic improvements to be noted. For example, replacing (8.1.15), one has

$$u \in C^2_\varphi \Longrightarrow A_j(v; y, x, \xi) \in C_\varphi S^1_{1,0} \cap S^1_{1,1}, \tag{9.2.2}$$

and replacing (8.1.10) there is

$$u \in C^2_\varphi \Longrightarrow (A_j - B_j)\langle\xi\rangle^{-1} \to 0 \text{ as } |\xi| \to \infty, \tag{9.2.3}$$

with a similar result for $K - \tilde{K}$, and for $A - \tilde{A}$ in (8.1.18). Furthermore, when you do symbol smoothing, the good parts belong to $S^m_{1,\delta}$, as before, and the remainder terms, such as K^b, arising in (8.1.19), have properties analogous to that of $M^b(x, D)$ in Lemma 9.1.A. Therefore there arises the following improvement of Proposition 8.1.A.

PROPOSITION 9.2.A. *If u satisfies the elliptic boundary problem (9.2.1), assumed to be regular, then*

$$\|u\|^2_{k+1,s-\frac{1}{2}} \le C(\|u\|_{C^2_\varphi})\Big[\sum_j \|f_j\|^2_{H^{s+k+1-\ell_j}(X)} + \|f\|^2_{k-1,s-\frac{1}{2}} + 1\Big] \tag{9.2.4}$$

provided $k = 0, 1, 2, \ldots,$ $u \in C^2_\varphi \cap H_{2,s-\frac{1}{2}}$, and $s > \frac{1}{2}$. Here, ℓ_j is the order of G_j in (9.2.1); $\ell_j = 0$ or 1.

Using the technique of §9.1, we can extend this result to the case when $\overline{M}$ is a general compact manifold, with boundary $\partial M = X$.

Such an estimate can be useful to implement the continuity method to solve nonlinear elliptic boundary problems. Thus, suppose we have a one parameter family of elliptic boundary problems, such as

$$F_\tau(x, D^2u) = f_\tau \text{ on } M, \quad G_\tau(u) = g_\tau \text{ on } \partial M, \tag{9.2.5}$$

where G_τ has order $\ell = 0$ or 1. We assume given $f_\tau \in C^\infty(\overline{M})$, $g_\tau \in C^\infty(\partial M)$.

PROPOSITION 9.2.B. *Suppose that for each $\tau \in [0, 1]$, smooth solutions to (9.2.5) satisfy a uniform a priori estimate in the C^2_φ-norm, for some modulus of continuity φ. Assume that for each τ, the linearization of (9.2.5) at $u \in C^\infty(\overline{M})$ is uniquely solvable. If (9.2.5) has a solution $u \in C^\infty(\overline{M})$ for $\tau = 0$, then it has a smooth solution for $\tau = 1$.*

PROOF: The method of continuity works. If J is the largest subinterval of $[0, 1]$, containing 0, such that (9.2.5) is solvable for $u \in C^\infty(\overline{M})$ when $\tau \in J$, the hypothesis on the linearized problems implies that J is open, by the implicit function theorem. The hypothesized bound on $\|u_\tau\|_{C^2_\varphi}$ together with the estimates (9.2.4) and the regularity theory of §8.1 imply that J is closed.

One of the most incisive uses of such a result appears in [CNS], on the solution to a Monge-Ampere equation

$$\begin{aligned} \det(\partial^2 u) &= f(x) \text{ in } M, \\ u &= g \text{ on } \partial M, \end{aligned} \tag{9.2.6}$$

given $M \subset \mathbb{R}^n$ strictly convex, $f \in C^\infty(\overline{M})$, $f > 0$, $g \in C^\infty(\partial M)$. The main result of [CNS] is:

THEOREM 9.2.C. *There exists a unique strictly convex solution $u \in C^\infty(\overline{M})$ of (9.2.6).*

The authors use the following case of (9.2.5). Construct a strictly convex $u_0 \in C^\infty(\overline{M})$ such that $u_0 = g$ on ∂M and $f_0 = \det(\partial^2 u_0) \geq f$ in M. Then for (9.2.5) take the family of boundary problems

$$\det(\partial^2 u) = \tau f + (1-\tau) f_0 \text{ on } M, \quad u = g \text{ on } \partial M. \tag{9.2.7}$$

This is clearly solvable at $\tau = 0$, and the linearization is well behaved. The task accomplished in [CNS] was to produce an estimate on the C^2_φ-norm of solutions to (9.2.7), with

$$\varphi(s) = \left[\log \frac{2}{s}\right]^{-1}, \quad 0 < s \leq 1. \tag{9.2.8}$$

This estimate then yields the existence result of Theorem 9.2.C.

A: Function spaces

§A.1: Hölder spaces, Zygmund spaces, and Sobolev spaces

Here we collect a few facts about various function spaces used in the paper. More details can be found in [S1], [Tr], [H4].

If $0 < s < 1$, we define $C^s(\mathbb{R}^n)$ to consist of bounded functions u such that

$$|u(x+y) - u(x)| \le C|y|^s. \tag{A.1.1}$$

For $k = 0, 1, 2, \dots$, we take $C^k(\mathbb{R}^n)$ to consist of bounded continuous functions u such that $D^\beta u$ is bounded and continuous, for $|\beta| \le k$. If $s = k + r$, $0 < r < 1$, we define $C^s(\mathbb{R}^n)$ to consist of functions $u \in C^k(\mathbb{R}^n)$ such that, for $|\beta| = k$, $D^\beta u$ belongs to $C^r(\mathbb{R}^n)$.

To connect Hölder spaces to Zygmund spaces, we use the S^0_1 partition of unity introduced in §1.3, $1 = \sum_{j=0}^{\infty} \psi_j(\xi)$ with ψ_j supported on $\langle \xi \rangle \sim 2^j$, and $\psi_j(\xi) = \psi_1(2^{1-j}\xi)$ for $j \ge 1$. It is known that, if $u \in C^s$, then

$$\sup_k 2^{ks} \|\psi_k(D)u\|_{L^\infty} < \infty. \tag{A.1.2}$$

To see this, first note that it is obvious for $s = 0$. For $s = \ell \in \mathbb{Z}^+$ it then follows from the elementary estimate

$$C_1 2^{k\ell} |\psi_k(D)u(x)| \le \sum_{|\alpha| \le \ell} |\psi_k(D) D^\alpha u(x)| \le C_2 2^{k\ell} |\psi_k(D)u(x)|. \tag{A.1.3}$$

Thus it suffices to establish that $u \in C^s$ implies (A.1.2) for $0 < s < 1$. Since $\hat{\psi}_1(x)$ has zero integral, we have, for $k \ge 1$,

$$\begin{aligned} |\psi_k(D)u(x)| &= \Big| \int \hat{\psi}_k(y) \big[u(x-y) - u(x) \big] \, dy \Big| \\ &\le C \int |\hat{\psi}_k(y)| \cdot |y|^s \, dy, \end{aligned} \tag{A.1.4}$$

which is readily bounded by $C\, 2^{-ks}$.

Conversely, if s is not an integer, finiteness in (A.1.2) implies $u \in C^s$. It suffices to demonstrate this for $0 < s < 1$. With $\Psi_k(\xi) = \sum_{j \le k} \psi_j(\xi)$, if $|y| \sim 2^{-k}$, write

$$\begin{aligned} u(x+y) - u(x) = &\int_0^1 y \cdot \nabla \Psi_k(D) u(x+ty) dt \\ &+ \big(I - \Psi_k(D)\big)\big(u(x+y) - u(x)\big) \end{aligned} \tag{A.1.5}$$

and use (A.1.2)-(A.1.3) to dominate the L^∞ norm of both terms on the right by $C \cdot 2^{-sk}$, since $\|\nabla \Psi_k(D)u\|_{L^\infty} \le C \cdot 2^{(1-s)k}$.

This converse breaks down if $s \in \mathbb{Z}^+$. We define the Zygmund class C_*^s to consist of u such that (A.1.2) is finite, using that to define the C_*^s-norm. Thus

$$C^s = C_*^s \text{ if } s \in \mathbb{R}^+ \setminus \mathbb{Z}^+, \quad C^k \subset C_*^k, \ k \in \mathbb{Z}^+. \tag{A.1.6}$$

It is known that C_*^s is an algebra for each $s > 0$. Also,

$$P \in OPS_{1,0}^m \Longrightarrow P : C_*^s \longrightarrow C_*^{s-m} \tag{A.1.7}$$

if $s, s-m > 0$. This is proved in §2.1. In fact, one can define C_*^s by finiteness of (A.1.2) for all $s \in \mathbb{R}$, and then (A.1.7) holds without restriction. In particular, with $\Lambda = (1-\Delta)^{\frac{1}{2}}$,

$$\Lambda^m : C_*^s \longrightarrow C_*^{s-m} \text{ is an isomorphism.} \tag{A.1.8}$$

The basic case C_*^1 can be characterized as the set of bounded continuous u such that

$$|u(x+y) + u(x-y) - 2u(x)| \le K|y|. \tag{A.1.9}$$

This is Zygmund's original definition.

For $1 < p < \infty$, $s \in \mathbb{R}$, the Sobolev spaces $H^{s,p}(\mathbb{R}^n)$ can be defined as

$$H^{s,p}(\mathbb{R}^n) = \Lambda^{-s}\big(L^p(\mathbb{R}^n)\big). \tag{A.1.10}$$

It is then true that, for $s = k \in \mathbb{Z}^+$, $u \in \mathcal{S}'(\mathbb{R}^n)$,

$$u \in H^{k,p}(\mathbb{R}^n) \Longleftrightarrow D^\alpha u \in L^p(\mathbb{R}^n) \text{ for } |\alpha| \le k. \tag{A.1.11}$$

There is a natural duality

$$H^{s,p}(\mathbb{R}^n)^* \approx H^{-s,p'}(\mathbb{R}^n), \quad \frac{1}{p} + \frac{1}{p'} = 1. \tag{A.1.12}$$

There is a characterization of $H^{s,p}(\mathbb{R}^n)$ analogous to (A.1.2), namely that

$$\Big\| \Big\{ \sum_{k=0}^{\infty} 4^{ks} |\psi_k(D)u|^2 \Big\}^{\frac{1}{2}} \Big\|_{L^p} \tag{A.1.13}$$

be finite. This is a consequence of the Littlewood-Paley theory of L^p. In fact, for $s = 0$, the equivalence of the norm above with $\|u\|_{L^p}$ is established in §0.11; see (0.11.32). For more general s, it follows from the simple estimate

$$C_1 2^{ks}|\psi_k(D)u(x)| \le |\psi_k(D)\Lambda^s u(x)| \le C_2 2^{ks}|\psi_k(D)u(x)|. \tag{A.1.14}$$

A complementary result, used in Lemma 2.1.G, is that, with $\Psi_k(\xi) = \sum\limits_{\ell\le k} \psi_\ell(\xi)$,

$$\Big\|\sum_{k=0}^{\infty} \Psi_k(D) f_k\Big\|_{H^{s,p}} \le C_{sp}\Big\|\Big\{\sum_{k=0}^{\infty} 4^{ks}|f_k|^2\Big\}^{\frac{1}{2}}\Big\|_{L^p}, \quad s > 0, \tag{A.1.15}$$

for $1 < p < \infty$. To prove this, writing $f_k = \Lambda^{-s}u_k$ and $\Psi_k(D)u_k = \sum\limits_{\ell=0}^{k} \psi_\ell(D)u_k$, the Littlewood-Paley estimates show that the left side of (A.1.15) is

$$\approx \Big\|\Big\{\sum_{\ell=0}^{\infty}\Big|\psi_\ell(D)\sum_{k=\ell}^{\infty} u_k\Big|^2\Big\}^{\frac{1}{2}}\Big\|_{L^p}, \tag{A.1.16}$$

which by (A.1.14) is

$$\approx \Big\|\Big\{\sum_{\ell=0}^{\infty} 4^{\ell s}\Big|\psi_\ell(D)\sum_{k=\ell}^{\infty} f_k\Big|^2\Big\}^{\frac{1}{2}}\Big\|_{L^p}. \tag{A.1.17}$$

Thus, with $w_k = 2^{ks}f_k$, we need to show that

$$\Big\|\Big\{\sum_{\ell=0}^{\infty} 4^{\ell s}\Big|\psi_\ell(D)\sum_{k=\ell}^{\infty} 2^{-ks}w_k\Big|^2\Big\}^{\frac{1}{2}}\Big\|_{L^p} \le C_{sp}\Big\|\Big\{\sum_{k=0}^{\infty} |w_k|^2\Big\}^{\frac{1}{2}}\Big\|_{L^p}; \tag{A.1.18}$$

in other words, we need so show continuity of

$$\Gamma(D) : L^p(\mathbb{R}^n, \ell^2) \longrightarrow L^p(\mathbb{R}^n, \ell^2) \tag{A.1.19}$$

where

$$\begin{aligned} \Gamma_{k\ell}(\xi) = \psi_k(\xi)2^{-(\ell-k)s}&, \quad \ell \ge k \\ 0 \qquad &, \quad \ell < k \end{aligned} \tag{A.1.20}$$

It is straightforward to verify that

$$\begin{aligned} \sum_k |D_\xi^\alpha \Gamma_{k\ell}(\xi)| &\le C\langle\xi\rangle^{-|\alpha|}, \quad s \ge 0, \\ \sum_\ell |D_\xi^\alpha \Gamma_{k\ell}(\xi)| &\le C_s\langle\xi\rangle^{-|\alpha|}, \quad s > 0, \end{aligned} \tag{A.1.21}$$

so

$$\|D_\xi^\alpha \Gamma(\xi)\|_{\mathcal{L}(\ell^2)} \le C_s \langle \xi \rangle^{-|\alpha|}, \quad s > 0. \tag{A.1.22}$$

Hence the vector valued Fourier multiplier result, Proposition 0.11.F, yields (A.1.19), and completes the proof of (A.1.15).

There are a number of useful 'Sobolev imbedding theorems.' We mention

$$H^{s,p}(\mathbb{R}^n) \subset L^{np/(n-sp)}(\mathbb{R}^n), \quad s < n/p. \tag{A.1.23}$$

Also

$$H^{s,p}(\mathbb{R}^n) \subset L^\infty(\mathbb{R}^n), \quad s > n/p. \tag{A.1.24}$$

In fact, (A.1.24) can be sharpened and extended to

$$H^{s,p}(\mathbb{R}^n) \subset C_*^r(\mathbb{R}^n), \quad r = s - n/p, \tag{A.1.25}$$

valid for all $s \in \mathbb{R}$.

§A.2. Morrey spaces

Let $p \in (1,\infty)$ and define $s \in (1-n, 1)$ by

$$1 - s = n/p. \tag{A.2.1}$$

By definition, the Morrey space $M^p(\mathbb{R}^n)$ consists of $f \in L^1(\mathbb{R}^n)$ such that

$$\int_{B_R} |f(x)|dx \le C\ R^{n-1+s} \tag{A.2.2}$$

for any ball B_R of radius $R \le 1$ centered at any point $y \in \mathbb{R}^n$. If Ω is a subset of $\mathbb{R}^n$, we say $u \in M^p(\Omega)$ if its extension by 0 outside Ω belongs to $M^p(\mathbb{R}^n)$. The notation involving p arises from the fact that, for Ω bounded,

$$L^p(\Omega) \subset M^p(\Omega), \tag{A.2.3}$$

as a consequence of Hölder's inequality. The role of Morrey spaces depends on the following result of Morrey.

THEOREM A.2.A. *If $\overline{\Omega}$ is smooth and bounded, $u \in H^{1,1}(\Omega)$, then*

$$\nabla u \in M^p(\Omega),\ p > n \Longrightarrow u \in C^s(\overline{\Omega}) \tag{A.2.4}$$

where p and s are related by (A.2.1).

In light of (A.2.3), this result is a bit sharper than the Sobolev imbedding theorem

$$H^{1,p}(\Omega) \subset C^s(\overline{\Omega}) \text{ for } s = 1 - n/p,\ p > n. \tag{A.2.5}$$

By taking an 'even reflection' of u across $\partial\Omega$, it suffices to prove (A.2.4) for $\Omega = \mathbb{R}^n$, u having compact support. This can be done as a consequence of the following results.

LEMMA A.2.B. *Let* $p(\lambda) = e^{-\lambda^2}$, $f \in L^1_{\text{comp}}(\mathbb{R}^n)$. *Then*

$$f \in M^p(\mathbb{R}^n) \Longleftrightarrow p(r\sqrt{-\Delta})|f| \le C\, r^{-1+s}, \quad r \in (0,1]. \tag{A.2.6}$$

PROOF: Exercise.

PROPOSITION A.2.C. *If* $g \in \mathcal{E}'(\mathbb{R}^n)$, $0 < s < 1$, *then*

$$g \in C^s(\mathbb{R}^n) \Longleftrightarrow \|p(r\sqrt{-\Delta})(\sqrt{-\Delta}\ g)\|_{L^\infty} \le c\, r^{-1+s}, \quad r \in (0,1]. \tag{A.2.7}$$

PROOF: This follows easily from the characterization (A.1.2) of $C^s(\mathbb{R}^n)$.

COROLLARY A.2.D. *If* $f \in L^1_{\text{comp}}(\mathbb{R}^n)$, *then*

$$f \in M^p(\mathbb{R}^n),\ p > n \Longrightarrow (-\Delta)^{-\frac{1}{2}} f \in C^s. \tag{A.2.8}$$

PROOF: $p(r\sqrt{-\Delta})\big((-\Delta)^{\frac{1}{2}}(-\Delta)^{-\frac{1}{2}}f\big) = p(r\sqrt{-\Delta})f$, and $|p(r\sqrt{-\Delta})f| \le p(r\sqrt{-\Delta})|f|$. Hence, if $f \in M^p(\mathbb{R}^n)$, the criterion (A.2.7) applies to $g = (-\Delta)^{-\frac{1}{2}}f$.

Now, to prove Morrey's Theorem, note that

$$\begin{aligned} \nabla u \in M^p(\mathbb{R}^n) &\Longrightarrow \nabla(-\Delta)^{-\frac{1}{2}} u \in C^s \\ &\Longrightarrow u \in C^s. \end{aligned} \tag{A.2.9}$$

It is useful to note that, given $q \in (1,\infty)$, $f \in L^q_{\text{comp}}(\mathbb{R}^n)$, $s = 1 - \frac{n}{p} \in (0,1)$,

$$\int_{B_R} |f(x)|^q dx \le C\ R^{n-q+qs} \Longrightarrow f \in M^p(\mathbb{R}^n). \tag{A.2.10}$$

Indeed, this follows easily from Hölder's inequality. We denote by $M^p_q(\mathbb{R}^n)$ the space of functions satisfying (A.2.10).

§A.3. BMO

Given $f \in L^1_{\text{loc}}(\mathbb{R}^n)$, and a cube Q with sides parallel to the coordinate axes, we set

$$f_Q = |Q|^{-1} \int_Q f(x)dx, \quad |Q| = \text{vol}(Q), \tag{A.3.1}$$

and say $f \in BMO$ if and only if

$$\sup_Q |Q|^{-1} \int_Q |f(x) - f_Q| dx = \|f\|_{BMO} \tag{A.3.2}$$

is finite. A constant has BMO-seminorm 0; (A.3.2) defines a norm on BMO modulo constants. Clearly $L^\infty \subset BMO$. It was shown by John and Nirenberg [JN] that, for a cube Q such as above,

$$\text{meas}\,\{x \in Q : |f(x) - f_Q| > \lambda\} \leq C\, e^{-c\lambda/\|f\|_{BMO}}. \tag{A.3.3}$$

Also, for any $p < \infty$,

$$|Q|^{-1} \int_Q |f(x) - f_Q|^p dx \leq C_p \|f\|^p_{BMO}. \tag{A.3.4}$$

Fefferman and Stein [FS] established many important results on BMO. We mention a few here. Define

$$f^\#(x) = \sup\{|Q|^{-1} \int_Q |f(x) - f_Q| dx : Q \ni x\}, \tag{A.3.5}$$

the sup being over all cubes containing x. Then, for $p \in (1, \infty)$,

$$\|f^\#\|_{L^p} \leq C_p \|f\|_{L^p}. \tag{A.3.6}$$

Of course, $f^\# \in L^\infty$ if and only if $f \in BMO$. Another very important result of [FS] is that f belongs to BMO if and only if it can be written in the form

$$f = g_0 + \sum_{j=1}^n R_j g_j, \quad g_j \in L^\infty(\mathbb{R}^n), \tag{A.3.7}$$

where R_j are the Riesz transforms; $(R_j g)\hat{}(\xi) = (\xi_j/|\xi|)\hat{g}(\xi)$.

Another important result in [FS] involves a connection between the space $BMO(\mathbb{R}^n)$ and Carleson measures on $\mathbb{R}^{n+1}_+$. If Q is a cube in $\mathbb{R}^n$, set

$$T(Q) = \{(x,y) \in \mathbb{R}^{n+1}_+ : x \in Q,\ 0 < y \le \ell(Q)\} \tag{A.3.8}$$

where $\ell(Q)$ is the length of a side of Q. Then a positive Borel measure on $\mathbb{R}^{n+1}_+$ is called a *Carleson measure* provided that, for all cubes $Q \subset \mathbb{R}^n$,

$$\mu(T(Q)) \le C|Q|. \tag{A.3.9}$$

The Carleson norm is

$$\|\mu\|_C = \sup_Q |Q|^{-1}\mu(T(Q)). \tag{A.3.10}$$

Let $\psi \in \mathcal{S}(\mathbb{R}^n)$, $\psi(0) = 0$. Given $f \in \mathcal{S}'(\mathbb{R}^n)$, $y > 0$, set $u(y,x) = \psi(yD)f(x)$. It is shown in [FS] that, if $f \in BMO$, then $|u(x,y)|^2 y^{-1} dx\, dy$ is a Carleson measure, and

$$\big\| |u|^2 y^{-1} dx\, dy \big\|_C \le C\|f\|^2_{BMO}. \tag{A.3.11}$$

We sketch a proof of this result, which will be used in Appendix D.

If $Q \subset \mathbb{R}^n$ is a cube, let $2Q$ denote the concentric cube with twice the diameter of Q. Since $\psi(yD)$ annihilates constants, we can alter f so that $\int_{2Q} f(x)dx = 0$. Write $f = f_0 + f_1$ where f_0 is the restriction of f to $2Q$. Then

$$\begin{aligned}\int_{T(Q)} |\psi(yD)f_0(x)|^2 y^{-1} dx\, dy &\le \int_{\mathbb{R}^{n+1}_+} |\psi(yD)f_0|^2 y^{-1} dx\, dy \\ &\le C\|f_0\|^2_{L^2}\end{aligned} \tag{A.3.12}$$

since one has the simple general estimate

$$\int_0^\infty \|\psi(yD)v\|^2_{L^2} dy/y \le C\|v\|^2_{L^2}. \tag{A.3.13}$$

Now $\|f_0\|^2_{L^2}$ is equal to

$$\int_{2Q} |f(x) - f_{2Q}|^2 dx \le C\|f\|^2_{BMO}|Q| \tag{A.3.14}$$

by (A.3.4). On the other hand, using (A.3.3), it is not hard to show that

$$|\psi(yD)f_1(x)| \le C\|f\|_{BMO}(y/\ell(Q)), \quad x \in Q, \tag{A.3.15}$$

which together with (A.3.12), (A.3.14) yields the desired estimate (A.3.11) on the integral of $|\psi(yD)f(x)|^2$ over $T(Q)$.

Finally we mention complements to (A.1.24) and (A.2.4), namely

$$\nabla u \in M^n(\mathbb{R}^n) \Longrightarrow u \in BMO, \tag{A.3.16}$$

and

$$H^{n/p,p}(\mathbb{R}^n) \subset BMO, \quad 1 < p < \infty. \tag{A.3.17}$$

B: Sup norm estimates

§B.1. L^∞ estimates on pseudodifferential operators

As is well known, an operator $P \in OPS^0_{1,0}$ need not be bounded on L^∞. We will establish a number of results on $\|Pu\|_{L^\infty}$, starting with the following, which is essentially given in the appendix to [BKM].

PROPOSITION B.1.A. *If $P \in OPS^0_{1,0}, s > n/2$, then*

$$\|Pu\|_{L^\infty} \le C\|u\|_{L^\infty} \cdot \Big[1 + \log\big(\|u\|_{H^s}/\|u\|_{L^\infty}\big)\Big]. \tag{B.1.1}$$

We suppose the norms are arranged to satisfy $\|u\|_{L^\infty} \le \|u\|_{H^s}$. Another way to write the result is in the form

$$\|Pu\|_{L^\infty} \le C\varepsilon^\delta\|u\|_{H^s} + C\Big(\log \frac{1}{\varepsilon}\Big)\|u\|_{L^\infty}, \tag{B.1.2}$$

for $0 < \varepsilon \le 1$, with C independent of ε. Then, letting $\varepsilon^\delta = \|u\|_{L^\infty}/\|u\|_{H^s}$ yields (B.1.1). The estimate (B.1.2) is valid when $s > \frac{1}{2}n + \delta$. This can be proved by writing $P = P_1 + P_2$, with $P_1 = P\Psi_1(\varepsilon D), \Psi_1 = \psi_0$ as in (1.3.1), and showing that

$$\|P_1 u\|_{L^\infty} \le C\Big(\log \frac{1}{\varepsilon}\Big)\|u\|_{L^\infty}, \quad \|P_2 u\|_{H^s} \le C\varepsilon^\delta\|u\|_{H^s}. \tag{B.1.3}$$

Rather than include the details on (B.1.3), we will derive (B.1.2) from an estimate relating the L^∞, H^s, and C^0_* norms, which has further uses.

It suffices to prove (B.1.2) with P replaced by $P+cI$, where c is greater than the L^2-operator norm of P; hence we can assume $P \in OPS^0_{1,0}$ is elliptic and invertible, with inverse $Q \in OPS^0_{1,0}$. Then (B.1.2) is equivalent to

$$\|u\|_{L^\infty} \le C\varepsilon^\delta\|u\|_{H^s} + C\Big(\log \frac{1}{\varepsilon}\Big)\|Qu\|_{L^\infty}.$$

Now since $Q : C^0_* \to C^0_*$, with inverse P, and the C^0_*-norm is weaker than the L^∞-norm, this estimate is a consequence of the following result.

PROPOSITION B.1.B. *If $s > \frac{1}{2}n + \delta$, then*

$$\|u\|_{L^\infty} \le C\varepsilon^\delta\|u\|_{H^s} + C\Big(\log \frac{1}{\varepsilon}\Big)\|u\|_{C^0_*}. \tag{B.1.4}$$

PROOF: Recall from (A.1.2) that $\|u\|_{C^0_*} = \sup_{j\ge 0} \|\psi_j(D)u\|_{L^\infty}$. Now, with $\Psi_j = \sum_{\ell\le j} \psi_\ell$, write $u = \Psi_j(D)u + (1 - \Psi_j(D))u$; let $\varepsilon = 2^{-j}$. Clearly

$$\|\Psi_j(D)u\|_{L^\infty} \le j\|u\|_{C^0_*}. \tag{B.1.5}$$

Meanwhile, using the Sobolev imbedding theorem, since $\frac{1}{2}n < s - \delta$,

$$\begin{aligned}\|(1 - \Psi_j(D))u\|_{L^\infty} &\leq C\|(1 - \Psi_j(D))u\|_{H^{s-\delta}} \\ &\leq C\, 2^{-j\delta}\|(1 - \Psi_j(D))u\|_{H^s},\end{aligned} \tag{B.1.6}$$

the last estimate holding since

$$\{2^{j\delta}\Lambda^{-\delta}(1 - \Psi_j(D)) : j \in \mathbb{Z}^+\} \text{ is bounded in } OPS^0_{1,0}. \tag{B.1.7}$$

The same reasoning shows that, if $1 < p < \infty$,

$$\|u\|_{L^\infty} \leq C\varepsilon^\delta \|u\|_{H^{s,p}} + C\Big(\log \frac{1}{\varepsilon}\Big)\|u\|_{C^0_*} \text{ if } s > \frac{n}{p} + \delta. \tag{B.1.8}$$

Note also that the arguments involving P in the proof of Proposition B.1.A work for $P \in OPS^0_{1,\delta}$, $0 \leq \delta < 1$. Hence Proposition B.1.A can be generalized and sharpened to:

PROPOSITION B.1.C. *If $P \in OPS^0_{1,\delta}$, $0 \leq \delta < 1$, and if $1 < p < \infty$, $s > n/p$, then*

$$\|Pu\|_{L^\infty} \leq C\|u\|_{C^0_*} \cdot \Big[1 + \log\big(\|u\|_{H^{s,p}}/\|u\|_{C^0_*}\big)\Big]. \tag{B.1.9}$$

The estimates (B.1.8)-(B.1.9) complement estimates of Brezis-Gallouet-Wainger [BrG], [BrW], which can be stated in the form

$$\|u\|_{L^\infty} \leq C\varepsilon^\delta \|u\|_{H^{s,p}} + C\Big(\log \frac{1}{\varepsilon}\Big)^{1-1/q} \|u\|_{H^{n/q,q}} \tag{B.1.10}$$

given

$$s > \frac{n}{p} + \delta, \quad q \in [2, \infty), \tag{B.1.11}$$

and a similar estimate for $q \in (1, 2]$, using $\big(\log\ 1/\varepsilon\big)^{1/q}$. This has a proof similar to that of (B.1.4) and (B.1.8). One uses instead of (B.1.5) the estimate

$$\|\Psi_j(D)u\|_{L^\infty} \leq Cj^{1-1/q}\|\Lambda^{n/q}u\|_{L^q}, \tag{B.1.12}$$

in case $2 \leq q < \infty$, and the analogous estimate for $1 < q \leq 2$.

The estimate (B.1.10) for $p = q = 2$ was used in [BrG] to produce a global existence result for a 'nonlinear Schrödinger equation.' Its role was to provide an energy estimate for which Gronwall's inequality would provide global bounds, in a fashion similar to [BKM] and to (5.3.18).

§B.2. The spaces $C^r_\#$

As stated at the end of §4.1, we are going to say

$$u \in C^0_\# \Longleftrightarrow \sup_\ell \langle \ell \rangle^{-K} \|\omega_\ell(D)u\|_{C^0} < \infty \tag{B.2.1}$$

for an exponent K which will be taken to be sufficiently large, as discussed below. The operators $\omega_\ell(D) \in OPS^0_{cl}$ have symbols $\omega_\ell(\xi/|\xi|)$ as in (1.1.16). Thus the ω_ℓ are the eigenfunctions of the Laplace operator on the sphere. Let us normalize them each to have L^∞-norm 1. Since $-\Delta_S\omega_\ell = \lambda_\ell^2\omega_\ell$ with $\lambda_\ell \sim C\ell^n$, it follows easily that

$$\|\omega_\ell\|_{C^k(S^{n-1})} \le C_{k\delta}\langle \ell \rangle^{kn+\delta} \tag{B.2.2}$$

for any $\delta > 0$, an estimate that has been implicitly used in Proposition 1.1.A. Using the decomposition

$$p(x,\xi) = \sum p_\ell(x)\omega_\ell(\xi/|\xi|) \tag{B.2.3}$$

we see that $p(x,\xi) \in C^0S^0_{cl}$, homogeneous in ξ, implies $\|p_\ell\|_{C^0}$ is rapidly decreasing. Hence, for any choice of $K \ge 0$ in (B.2.1), it is clear that

$$p(x,\xi) \in C^0S^0_{cl} \Longrightarrow p(x,D) : C^0_\# \longrightarrow C^0. \tag{B.2.4}$$

Now Proposition B.1.A implies that, if $s > \frac{1}{2}n + \delta$,

$$\|\omega_\ell(D)u\|_{L^\infty} \le C_\ell\Big[\varepsilon^\delta\|u\|_{H^s} + \Big(\log\frac{1}{\varepsilon}\Big)\|u\|_{L^\infty}\Big]. \tag{B.2.5}$$

Furthermore, C_ℓ can be estimated in terms of finitely many derivatives of ω_ℓ. Therefore, there is a $K = K(n)$ such that

$$C_\ell \le C\langle \ell \rangle^K. \tag{B.2.6}$$

In (B.2.1) we pick K large enough that (B.2.6) holds. Then we obtain the estimate

$$\|u\|_{C^0_\#} \le C\varepsilon^\delta\|u\|_{H^s} + C\Big(\log\frac{1}{\varepsilon}\Big)\|u\|_{C^0}, \tag{B.2.7}$$

hence

$$\|u\|_{C^0_\#} \le C\|u\|_{C^0} \cdot \Big[1 + \log\big(\|u\|_{H^s}/\|u\|_{C^0}\big)\Big]. \tag{B.2.8}$$

In fact, using Proposition B.1.C, we obtain

$$\|u\|_{C^0_\#} \le C\|u\|_{C^0_*} \cdot \Big[1 + \log\big(\|u\|_{H^s}/\|u\|_{C^0_*}\big)\Big], \tag{B.2.9}$$

as long as $s > n/2$.

In (4.1.30) we defined

$$C^r_\# = \Lambda^{-r}(C^0_\#), \tag{B.2.10}$$

and noted that, for any $r \ge 0$,

$$P \in OPS^0_{cl} \Longrightarrow P : C^r_\# \longrightarrow C^r. \tag{B.2.11}$$

Now, replacing u by $\Lambda^r u$ in (B.2.9), we see that, for any $r \ge 0$,

$$\|u\|_{C^r_\#} \le C\|u\|_{C^r_*} \cdot \Big[1 + \log\big(\|u\|_{H^s}/\|u\|_{C^r_*}\big)\Big], \quad s > \frac{n}{2} + r. \tag{B.2.12}$$

This is in effect a useful extension of Proposition B.1.A. One consequence is the estimate (5.3.17), which plays an important role in the proof of Theorem 5.3.A.

C: DeGiorgi-Nash-Moser estimates

In this appendix we establish regularity for a class of PDE $Lu = f$, for second order operators of the form (using the summation convention)

$$Lu = b^{-1}\partial_j a^{jk} b\, \partial_k u \tag{C.0.1}$$

where $(a^{jk}(x))$ is a positive definite bounded matrix and $0 < b_0 \le b(x) \le b_1$, b scalar, and a^{jk}, b are merely measurable. We will present Moser's derivation of interior bounds and Hölder continuity of solutions to $Lu = 0$, from [Mo2], in §C.1-§3.2, and then Morrey's analysis of the inhomogeneous equation $Lu = f$ and proof of boundary regularity, in §C.3-§C.4, from [Mor]. Other proofs can be found in [GT] and [KS].

We make a few preliminary remarks on (C.0.1). We will use a^{jk} to define an inner product of vectors:

$$\langle V, W\rangle = V_j a^{jk} W_k, \tag{C.0.2}$$

and use $b\,dx = dV$ as the volume element. In case $g_{jk}(x)$ is a metric tensor, one can take $a^{jk} = g^{jk}$ and $b = g^{\frac{1}{2}}$; then (C.0.1) defines the Laplace operator. For compactly supported w,

$$(Lu, w) = -\int \langle \nabla u, \nabla w\rangle dV. \tag{C.0.3}$$

The behavior of L on a nonlinear function of u, $v = f(u)$, plays an important role in estimates; we have

$$v = f(u) \Longrightarrow Lv = f'(u)Lu + f''(u)|\nabla u|^2, \tag{C.0.4}$$

where we set $|V|^2 = \langle V, V\rangle$. Also, taking $w = \psi^2 u$ in (C.0.3) gives the following important identity. If $Lu = g$ on an open set Ω and $\psi \in C_0^1(\Omega)$, then

$$\int \psi^2 |\nabla u|^2 dV = -2\int \langle \psi\nabla u, u\nabla\psi\rangle dV - \int \psi^2 g u\, dV. \tag{C.0.5}$$

Compare (2.2.67). Applying Cauchy's inequality to the first term on the right yields the useful estimate

$$\frac{1}{2}\int \psi^2 |\nabla u|^2 dV \le 2\int |u|^2 |\nabla\psi|^2 dV - \int \psi^2 g u\, dV. \tag{C.0.6}$$

Given these preliminaries, we are ready to present Moser's analysis.

§C.1. Moser iteration and L^∞ estimates

Consider a nested sequence of open sets with smooth boundary

$$\Omega_0 \supset \cdots \supset \Omega_j \supset \Omega_{j+1} \supset \cdots \tag{C.1.1}$$

with intersection $\mathcal{O}$. We will make the geometrical hypothesis that the distance of any point on $\partial\Omega_{j+1}$ to $\partial\Omega_j$ is $\sim Cj^{-2}$. We want to estimate the sup norm of a function v on $\mathcal{O}$ in terms of its L^2-norm on Ω_0, assuming

$$v > 0 \text{ is a subsolution of } L, \text{ i.e., } Lv \geq 0. \tag{C.1.2}$$

In view of (C.0.4), an example is

$$v = (1+u^2)^{\frac{1}{2}}, \quad Lu = 0. \tag{C.1.3}$$

We will obtain such an estimate in terms of the Sobolev constants $\gamma(\Omega_j)$ and C_j, defined below. Ingredients for the analysis include the following two lemmas, the first being a standard Sobolev inequality.

LEMMA C.1.A. *For $v \in H^1(\Omega_j)$, $\kappa \leq n/(n-2)$,*

$$\|v^\kappa\|^2_{L^2(\Omega_j)} \leq \gamma(\Omega_j)\left[\|\nabla v\|^{2\kappa}_{L^2(\Omega_j)} + \|v\|^{2\kappa}_{L^2(\Omega_j)}\right]. \tag{C.1.4}$$

The next lemma follows from (C.0.6), if we take $\psi = 1$ on Ω_{j+1}, tending roughly linearly to 0 on $\partial\Omega_j$.

LEMMA C.1.B. *If $v > 0$ is a subsolution of L, then, with $C_j = C(\Omega_j, \Omega_{j+1})$,*

$$\|\nabla v\|_{L^2(\Omega_{j+1})} \leq C_j \|v\|_{L^2(\Omega_j)}. \tag{C.1.5}$$

Under the geometrical conditions indicated above on Ω_j, we can assume

$$\gamma(\Omega_j) \leq \gamma_0, \quad C_j \leq C(j^2+1). \tag{C.1.6}$$

Putting together the two lemmas, we see that, when v satisfies (C.1.2),

$$\begin{aligned} \|v^\kappa\|^2_{L^2(\Omega_{j+1})} &\leq \gamma(\Omega_{j+1})\left[C_j^{2\kappa}\|v\|^{2\kappa}_{L^2(\Omega_j)} + \|v\|^{2\kappa}_{L^2(\Omega_{j+1})}\right] \\ &\leq \gamma_0 (C_j^{2\kappa}+1)\|v\|^{2\kappa}_{L^2(\Omega_j)}. \end{aligned} \tag{C.1.7}$$

Fix $\kappa \in (1, n/(n-2)]$. Now, if v satisfies (C.1.2), so does

$$v_j = v^{\kappa^j}, \tag{C.1.8}$$

by (C.0.4). Note that $v_{j+1} = v_j^\kappa$. Now let

$$N_j = \|v\|_{L^{2\kappa^j}(\Omega_j)} = \|v_j\|_{L^2(\Omega_j)}^{1/\kappa^j}, \tag{C.1.9}$$

so

$$\|v\|_{L^\infty(\mathcal{O})} = \limsup_{j\to\infty} N_j. \tag{C.1.10}$$

If we apply (C.1.7) to v_j, we have

$$\|v_{j+1}\|^2_{L^2(\Omega_{j+1})} \le \gamma_0(C_j^{2\kappa} + 1)\|v_j\|^{2\kappa}_{L^2(\Omega_j)}. \tag{C.1.11}$$

Note that the left side is equal to $N_{j+1}^{2\kappa^{j+1}}$ and the norm on the right is equal to $N_j^{2\kappa^{j+1}}$. Thus (C.1.11) is equivalent to

$$N_{j+1}^2 \le \Big[\gamma_0(C_j^{2\kappa} + 1)\Big]^{1/\kappa^{j+1}} N_j^2. \tag{C.1.12}$$

By (C.1.6), $C_j^{2\kappa} + 1 \le C_0(j^{4\kappa} + 1)$, so

$$\begin{aligned}\limsup_{j\to\infty} N_j^2 &\le \prod_{j=0}^{\infty}\Big[\gamma_0 C_0(j^{4\kappa} + 1)\Big]^{1/\kappa^{j+1}} N_0^2 \\ &\le (\gamma_0 C_0)^{1/(\kappa-1)}\Big[\exp\sum_{j=0}^{\infty}\kappa^{-j-1}\ \log(j^{4\kappa}+1)\Big] N_0^2 \\ &\le K^2 N_0^2,\end{aligned} \tag{C.1.13}$$

for finite K. This gives Moser's sup norm estimate:

THEOREM C.1.C. *If $v > 0$ is a subsolution of L, then*

$$\|v\|_{L^\infty(\mathcal{O})} \le K\|v\|_{L^2(\Omega_0)} \tag{C.1.14}$$

where $K = K(\gamma_0, C_0, n)$.

§C.2. Hölder continuity

Hölder continuity of a solution to $Lu = 0$ is obtained as a consequence of the following Harnack inequality. Let $B_\rho = \{x : |x| < \rho\}$.

PROPOSITION C.2.A. *Let $u \ge 0$ be a solution of $Lu = 0$ in B_{2r}. Suppose*

$$\text{meas}\ \{x \in B_r : u(x) > 1\} > c_0^{-1} r^n. \tag{C.2.1}$$

Then there is a constant $c > 0$ such that

$$u(x) > c^{-1} \text{ in } B_{r/2}. \tag{C.2.2}$$

This will be established by examining $v = f(u)$ with

$$f(u) = \max\{-\log(u+\varepsilon), 0\}, \tag{C.2.3}$$

where ε is chosen in $(0,1)$. Note that f is convex, so v is a subsolution. Our first goal will be to estimate the $L^2(B_r)$-norm of ∇v. Once this is done, Theorem C.1.C will be applied to estimate v from above (hence u from below) on $B_{r/2}$.

We begin with a variant of (C.0.5), obtained by taking $w = \psi^2 f'(u)$ in (C.0.3). The identity is

$$\int \psi^2 f'' |\nabla u|^2 dV + 2\int \langle \psi f' \nabla u, \nabla \psi \rangle dV = -(Lu, \psi^2 f'). \tag{C.2.4}$$

This vanishes if $Lu = 0$. Note that $f'\nabla u = \nabla v$ and $f'^2 |\nabla u|^2 = |\nabla v|^2$, if $v = f(u)$. Applying Cauchy's inequality to the second integral, we obtain

$$\int \psi^2 \left[\frac{f''}{f'^2} - \delta^2\right] |\nabla v|^2 dV \le \frac{1}{\delta^2} \int |\nabla \psi|^2 dV. \tag{C.2.5}$$

Now the function $f(u)$ in (C.2.3) has the property that

$$h = -e^{-f} \text{ is a convex function;} \tag{C.2.6}$$

indeed, in this case $h(u) = \max\ \{-(u+\varepsilon), -1\}$. Thus

$$f'' - f'^2 = e^f h'' \ge 0. \tag{C.2.7}$$

Hence (C.2.5) yields (with $\delta^2 = \frac{1}{2}$)

$$\int \psi^2 |\nabla v|^2 dV \le 4 \int |\nabla \psi|^2 dV, \tag{C.2.8}$$

after one overcomes the minor problem that f' has a jump discontinuity. If we pick ψ to $= 1$ on B_r and go linearly to 0 on ∂B_{2r}, we obtain the estimate

$$\int_{B_r} |\nabla v|^2 dV \le C r^{n-2} \tag{C.2.9}$$

for $v = f(u)$, given that $Lu = 0$ and that (C.2.6) holds.

Now the hypothesis (C.2.1) implies that v vanishes on a subset of B_r of measure $> c_0^{-1} r^n$. Hence there is an elementary estimate of the form

$$r^{-n} \int_{B_r} v^2 dV \le C r^{2-n} \int_{B_r} |\nabla v|^2 dV, \tag{C.2.10}$$

which is bounded from above by (C.2.9). Now Theorem C.1.C, together with a simple scaling argument, gives

$$v(x)^2 \le C r^{-n} \int_{B_r} v^2 dV \le C_1^2, \quad x \in B_{r/2}, \tag{C.2.11}$$

so

$$u + \varepsilon \ge e^{-C_1} \text{ for } x \in B_{r/2}, \tag{C.2.12}$$

for all $\varepsilon \in (0,1)$. Taking $\varepsilon \to 0$, we have the proof of Proposition C.2.A.

We remark that Moser obtained a stronger Harnack inequality in [Mo3], by a more elaborate argument.

To deduce Hölder continuity of a solution to $Lu = 0$ given Proposition C.2.A is fairly simple. Following [Mo2], who followed DeGiorgi, we have from §C.1 a bound

$$|u(x)| \le K \tag{C.2.13}$$

on any compact subset $\mathcal{O}$ of Ω_0, given $u \in H^1(\Omega_0)$, $Lu = 0$. Fix $x_0 \in \mathcal{O}$, such that $B_\rho(x_0) \subset \mathcal{O}$, and, for $r \le \rho$, let

$$\omega(r) = \sup_{B_r} u(x) - \inf_{B_r} u(x), \tag{C.2.14}$$

where $B_r = B_r(x_0)$. Clearly $\omega(\rho) \le 2K$. Adding a constant to u, we can assume

$$\sup_{B_\rho} u(x) = -\inf_{B_\rho} u(x) = \frac{1}{2}\omega(\rho) = M. \tag{C.2.15}$$

Then $u_+ = 1 + u/M$ and $u_- = 1 - u/M$ are also annihilated by L. They are both ≥ 0 and at least one of them satisfies the hypothesis (C.2.1), with $r = \rho/2$. If for example u_+ does, then Proposition C.2.A implies

$$u_+(x) > c^{-1} \text{ in } B_{\rho/4}, \tag{C.2.16}$$

so

$$-M\Big(1-\frac{1}{c}\Big)\le u(x)\le M \text{ in } B_{\rho/4}. \tag{C.2.17}$$

Hence

$$\omega(\rho/4)\le\Big(1-\frac{1}{2c}\Big)\omega(\rho), \tag{C.2.18}$$

which gives Hölder continuity:

$$\omega(r)\le\omega(\rho)\Big(\frac{r}{\rho}\Big)^{\alpha},\quad \alpha=-\log_4\Big(1-\frac{1}{2c}\Big). \tag{C.2.19}$$

We state the result formally.

THEOREM C.2.B. *If $u\in H^1(\Omega_0)$ solves $Lu=0$ then for every compact $\mathcal{O}$ in Ω_0, there is an estimate*

$$||u||_{C^{\alpha}(\mathcal{O})}\le C||u||_{L^2(\Omega_0)}. \tag{C.2.20}$$

In fact, we can show that $\nabla u|_{\mathcal{O}}$ belongs to the Morrey space M_2^p, with $p=n/(1-r)$, which is stronger than (C.2.20), by Theorem A.2.A. To see this, if B_R is a ball of radius R centered at y, $B_{2R}\subset\Omega$, then let $c=u(y)$ and replace u by $u(x)-c$ in (C.0.6), to get

$$\frac{1}{2}\int\psi^2|\nabla u|^2dV\le 2\int|u(x)-c|^2|\nabla\psi|^2dV. \tag{C.2.21}$$

Taking $\psi=1$ on B_R, going linearly to 0 on ∂B_{2R}, gives

$$\int_{B_R}|\nabla u|^2dV\le C\ R^{n-2+2r}, \tag{C.2.22}$$

as asserted. Compare (8.3.31).

§C.3. Inhomogeneous equations

We take L as in (C.0.1), with a^{jk} measurable, satisfying

$$0<\lambda_0|\xi|^2\le\sum a^{jk}(x)\xi_j\xi_k\le\lambda_1|\xi|^2 \tag{C.3.1}$$

while for simplicity we assume $b,b^{-1}\in\ \mathrm{Lip}(\overline{\Omega})$. We consider a PDE

$$Lu=f. \tag{C.3.2}$$

It is clear that, for $u \in H_0^1(\Omega)$,

$$(Lu, u) \geq C \sum \|\partial_j u\|_{L^2}^2, \tag{C.3.3}$$

so we have an isomorphism

$$L : H_0^1(\Omega) \xrightarrow{\approx} H^{-1}(\Omega). \tag{C.3.4}$$

Thus, for any $f \in H^{-1}(\Omega)$, (C.3.2) has a unique solution $u \in H_0^1(\Omega)$. One can write such f as

$$f = \sum \partial_j g_j, \quad g_j \in L^2(\Omega). \tag{C.3.5}$$

The solution $u \in H_0^1(\Omega)$ then satisfies

$$\|u\|_{H^1(\Omega)}^2 \leq C \sum \|g_j\|_{L^2}^2. \tag{C.3.6}$$

Here C depends on $\Omega, \lambda_0, \lambda_1$, and $b \in \text{Lip}(\overline{\Omega})$.

One can also consider the boundary problem

$$Lv = 0 \text{ on } \Omega, \quad v = w \text{ on } \partial\Omega, \tag{C.3.7}$$

given $w \in H^1(\Omega)$, where the latter condition means $v - w \in H_0^1(\Omega)$. Indeed, setting $v = u + w$, the equation for u is $Lu = -Lw$, $u \in H_0^1(\Omega)$. Thus (C.3.7) is uniquely solvable, with an estimate

$$\|\nabla v\|_{L^2(\Omega)} \leq C \|\nabla w\|_{L^2(\Omega)} \tag{C.3.8}$$

where C has a dependence as in (C.3.6).

The main goal of this section is to give Morrey's proof of the following local regularity result.

THEOREM C.3.A. *Suppose $u \in H^1(\Omega)$ solves (C.3.2), with $f = \sum \partial_j g_j$, $g_j \in M_2^q(\Omega)$, $q > n$, i.e.,*

$$\int_{B_r} |g_j|^2 dV \leq K_1^2 (r/R)^{n-2+2\mu}, \quad 0 < \mu < 1. \tag{C.3.9}$$

Then, for any $\mathcal{O} \subset\subset \Omega$, $u \in C^\mu(\mathcal{O})$. In fact

$$\int_{B_r} |\nabla u|^2 dV \leq K_2^2 (r/R)^{n-2+2\mu}. \tag{C.3.10}$$

Morrey established this by using (C.3.6), (C.3.8), and an elegant dilation argument, in concert with the results of §C.2. For this, suppose $B_R = B_R(y) \subset \Omega$ for each $y \in \mathcal{O}$. We can write $u = U + H$ on B_R where

$$\begin{aligned} LU &= \sum \partial_j g_j \text{ on } B_R, \quad U \in H_0^1(B_R), \\ LH &= 0 \text{ on } B_R, \quad H - u \in H_0^1(B_R), \end{aligned} \tag{C.3.11}$$

and we have

$$\|\nabla U\|_{L^2(B_R)} \le C_1 \|g\|_{L^2(B_R)}, \quad \|\nabla H\|_{L^2(B_R)} \le C_2 \|\nabla u\|_{L^2(B_R)}, \tag{C.3.12}$$

where $\|g\|_{L^2}^2 = \sum \|g_j\|_{L^2}^2$. Let us set

$$\|F\|_r = \|F\|_{L^2(B_r)}. \tag{C.3.13}$$

Also let $\kappa(g_j, R)$ be the best constant K_1 for which (C.3.9) is valid for $0 < r \le R$. If $g_\tau(x) = g(\tau x)$, note that

$$\kappa(g_\tau, S/\tau) = \tau^{n/2} \kappa(g, S).$$

Now define

$$\begin{aligned} \varphi(r) = \sup\{\|\nabla U\|_{rS} : U \in H_0^1(B_S),\ LU = \sum \partial_j g_j, \text{ on } B_S, \\ \kappa(g_j, S) \le 1, 0 < S \le R\}. \end{aligned} \tag{C.3.14}$$

Let us denote by $\varphi_S(r)$ the sup in (C.3.14) with S fixed, in $(0, R]$. Then $\varphi_S(r)$ coincides with $\varphi_R(r)$, with L replaced by the dilated operator, coming from the dilation taking B_S to B_R. More precisely, the dilated operator is

$$L_S = b_S\ \partial_j\ a_S^{jk}\ b_S^{-1}\ \partial_k,$$

with

$$a_S^{jk}(x) = a^{jk}(Sx/R), \quad b_S(x) = b(Sx/R),$$

assuming 0 has been arranged to be the center of B_R. To see this, note that, if $\tau = S/R$, $U_\tau(x) = \tau^{-1} U(\tau x)$, and $g_{j\tau}(x) = g_j(\tau x)$, then

$$LU = \sum \partial_j g_j \Longleftrightarrow L_S U_\tau = \sum \partial_j g_{j\tau}.$$

Also, $\nabla U_\tau(x) = (\nabla U)(\tau x)$, so $\|\nabla U_\tau\|_{S/\tau} = \tau^{n/2} \|\nabla U\|_S$.

Now for this family L_S, one has a *uniform* bound on C in (C.3.6); hence $\varphi(r)$ is *finite* for $r \in (0, 1]$. We also note that the bounds in (C.2.20) and (C.2.22) are uniformly valid for this family of operators. Theorem C.3.A will be proved when we show that

$$\varphi(r) \leq A\, r^{\frac{n}{2}-1+\mu}. \tag{C.3.15}$$

In fact, this will give the estimate (C.3.10) with u replaced by U; meanwhile such an estimate with u replaced by H is a consequence of (C.2.22). Let H satisfy (C.2.22) with $r = \mu_0$. We take $\mu < \mu_0$.

Pick $S \in (0, R]$ and pick g_j satisfying (C.3.9), with R replaced by S and K_1 by K. Write the U of (C.3.11) as $U = U_S + H_S$ on B_S, where $U_S \in H_0^1(B_S)$, $LU_S = LU = \sum \partial_j g_j$ on B_S. Clearly (C.3.9) implies

$$\int_{B_r} |g_j|^2 dV \leq K^2 (S/R)^{n-2+2\mu} (r/S)^{n-2+2\mu}. \tag{C.3.16}$$

Thus, as in (C.3.12) (and recalling the definition of φ), we have

$$\begin{aligned} \|\nabla U_S\|_S &\leq A_1 K (S/R)^{\frac{n}{2}-1+\mu}, \\ \|\nabla H_S\|_S &\leq A_2 \|\nabla U\|_S \leq A_2 K \varphi(S/R). \end{aligned} \tag{C.3.17}$$

Now, suppose $0 < r < S < R$. Then, applying (C.2.22) to H_S, we have

$$\begin{aligned} \|\nabla U\|_r &\leq \|\nabla U_S\|_r + \|\nabla H_S\|_r \\ &\leq K (S/R)^{\frac{n}{2}-1+\mu} \varphi(r/S) + A_3 K \varphi(S/R)\, (r/S)^{\frac{n}{2}-1+\mu_0}. \end{aligned} \tag{C.3.18}$$

Therefore, setting $s = r/R$, $t = S/R$, we have the inequality

$$\varphi(s) \leq t^{\frac{n}{2}-1+\mu} \varphi(s/t) + A_3 \varphi(t)\, (s/t)^{\frac{n}{2}-1+\mu_0}, \tag{C.3.19}$$

valid for $0 < s < t \leq 1$. Since it is clear that $\varphi(r)$ is monotone and finite on $(0, 1]$, it is an elementary exercise to deduce from (C.3.19) that $\varphi(r)$ satisfies an estimate of the form (C.3.15), as long as $\mu < \mu_0$. This proves Theorem C.3.A.

§C.4. Boundary regularity

Now that we have interior regularity estimates for the inhomogeneous problem, we will be able to use a few simple tricks to establish regularity up to the boundary for solutions to the Dirichlet problem

$$Lu = \sum \partial_j g_j, \quad u = f \text{ on } \partial\Omega, \tag{C.4.1}$$

where L has the form (C.0.1), $\overline{\Omega}$ is compact with smooth boundary, $f \in \text{Lip}(\partial\Omega)$, $g_j \in L^q(\Omega)$, $q > n$. First, extend f to $f \in \text{Lip}(\overline{\Omega})$. Then $u = v + f$ where v solves

$$Lv = \sum \partial_j h_j, \quad v = 0 \text{ on } \partial\Omega, \tag{C.4.2}$$

where

$$\partial_j h_j = \partial_j g_j - b^{-1} \partial_j a^{jk} b \, \partial_k f. \tag{C.4.3}$$

We will assume $b \in \text{Lip}(\overline{\Omega})$; then h_j can be chosen in L^q also.

The class of equations (C.4.2) is invariant under smooth changes of variables (indeed, invariant under Lipschitz homeomorphisms with Lipschitz inverses, having the further property of preserving volume up to a factor in $\text{Lip}(\overline{\Omega})$). Thus make a change of variables to flatten out the boundary (locally), so we consider a solution $v \in H^1$ to (C.4.2) in $x_n > 0$, $|x| \leq R$. We can even arrange that $b = 1$. Now extend v to negative x_n, to be *odd* under the reflection $x_n \mapsto -x_n$. Also extend $a^{jk}(x)$ to be even when $j, k < n$ or $j = k = m$, and odd when j or $k = n$ (but not both). Extend h_j to be odd for $j < n$ and even for $j = n$. With these extensions, we continue to have (C.4.2) holding, this time in the ball $|x| \leq R$. Thus interior regularity applies to this extension of v, yielding Hölder continuity. The following is hence proved.

THEOREM C.4.A. *For L of the form* (C.0.1), *elliptic, with $b \in Lip(\overline{\Omega})$ and $a^{jk} \in L^\infty(\Omega)$, a solution $u \in H^1(\Omega)$ to* (C.4.1), *with $g_j \in L^q(\Omega)$, $q > n$, $f \in Lip(\partial\Omega)$, has a Hölder estimate*

$$||u||_{C^\mu(\overline{\Omega})} \leq C\Big[\sum ||g_j||_{L^q(\Omega)} + ||f||_{Lip(\partial\Omega)}\Big]. \tag{C.4.4}$$

D: Paraproduct estimates

In §3.5 a paraproduct estimate of Coifman and Meyer, stated as Theorem 3.5.A, played an important role. We prove some related results here. For the sake of brevity, we establish a result which is a bit simpler than Theorem 3.5.A, but which nevertheless suffices for all the applications in §3.5. To begin, let $\varphi, \psi, \rho \in \mathcal{S}(\mathbb{R}^n)$, and set

$$P_t = \varphi(tD), \quad Q_t = \psi(tD), \quad R_t = \rho(tD). \tag{D.1.1}$$

PROPOSITION D.1.A. *If $\varphi, \psi, \rho \in \mathcal{S}(\mathbb{R}^n)$ and $\psi(0) = \rho(0) = 0$, then*

$$\tau(a, f) = \int_0^\infty R_t\big((Q_t a) \cdot (P_t f)\big) t^{-1} dt \tag{D.1.2}$$

satisfies an estimate

$$\|\tau(a, f)\|_{L^2} \le C\|a\|_{BMO}\|f\|_{L^2}. \tag{D.1.3}$$

Theorem 3.5.A dealt with a variant of (D.1.2), without the R_t.

To begin the proof of Proposition D.1.A, pick $g \in L^2(\mathbb{R}^n)$ and write

$$\begin{aligned} &\big(\tau(a, f), g\big) \\ &= \int_0^\infty \int (R_t^* g)(Q_t a)(P_t f)\ t^{-1} dx\ dt \\ &\le \Big(\int_0^\infty \int |R_t^* g(x)|^2 t^{-1} dx\ dt\Big)^{\frac{1}{2}} \cdot \Big(\int_0^\infty \int |Q_t a|^2 |P_t f|^2 t^{-1} dx\ dt\Big)^{\frac{1}{2}}. \end{aligned} \tag{D.1.4}$$

Given that $\rho(0) = 0$, it is easy to see that

$$\int_0^\infty \|R_t^* g\|_{L^2}^2\ t^{-1} dt \le C\|g\|_{L^2}^2 \tag{D.1.5}$$

so we bound the square of (D.1.4) by

$$C\|g\|_{L^2}^2 \int\int |P_t f(x)|^2 d\mu(t, x). \tag{D.1.6}$$

where

$$d\mu(t, x) = |Q_t a(x)|^2 t^{-1} dx\ dt. \tag{D.1.7}$$

Now, if $a \in BMO(\mathbb{R}^n)$, μ is a *Carleson measure*, i.e., for any cube $Q \subset \mathbb{R}^n$, of length $\ell(Q)$, if we set $T(Q) = \{(t,x) : x \in Q, 0 < t \le \ell(Q)\}$, then

$$\mu(T(Q)) \le K \ \mathrm{vol}(Q). \tag{D.1.8}$$

A proof of this is sketched in §A.3, where it is stated as (A.3.11). The best constant $K = ||\mu||_C$ in (D.1.9) satisfies

$$||\mu||_C \le C||a||^2_{BMO}. \tag{D.1.9}$$

To proceed, we have the following result of Carleson.

LEMMA D.1.B. *If μ is a Carleson measure on $\mathbb{R}^{n+1}_+$, then*

$$\int\int |P_t f(x)|^2 \ d\mu(x,t) \le C||\mu||_C \int_{\mathbb{R}^n} |Mf(x)|^2 \ dx, \tag{D.1.10}$$

where Mf is the Hardy-Littlewood maximal function.

The proof of the Lemma is essentially elementary. Using a Vitali covering argument one shows that, for all $\lambda > 0$,

$$\mu\{(x,t) : |P_t f(x)| > \lambda\} \le C \ \mathrm{meas} \ \{x \in \mathbb{R}^n : Mf(x) > \lambda/C\}. \tag{D.1.11}$$

It is classical that $||Mf||^2_{L^2} \le C||f||^2_{L^2}$, so with (D.1.10) in concert with (D.1.6), one has a bound

$$|(\tau(a,f),g)| \le C||a||_{BMO}||f||_{L^2}||g||_{L^2}, \tag{D.1.12}$$

establishing (D.1.3).

The following result will suffice to establish the estimate (3.5.18) on $R(a,f)$.

PROPOSITION D.1.C. *Let $\varphi, \psi \in C_0^\infty(\mathbb{R}^n)$ be supported on $0 < K \le |\xi| \le L$. Then, with P_t and Q_t as in* (D.1.1),

$$\tau(a,f) = \int_0^\infty (Q_t a) \cdot (P_t f) \ t^{-1} dt \tag{D.1.13}$$

satisfies the estimate (D.1.3).

PROOF: Pick $\rho \in C_0^\infty(\mathbb{R}^n)$, equal to 1 on $|\xi| \le 2L$. With $R_t = \rho(tD)$, we have

$$(Q_t a) \cdot (P_t f) = R_t\big((Q_t a) \cdot (P_t f)\big). \tag{D.1.14}$$

Thus (D.1.13) actually has the form (D.1.2). However, here we do not have $\rho(0) = 0$. But, looking at the first integral expression in (D.1.4) for $(\tau(a, f), g)$, we see that we can *interchange* the roles of f and g, and then the arguments leading to the estimate (D.1.12) go through.

Similarly, the following result will suffice to establish the estimate (3.5.19) on $T_f a$.

PROPOSITION D.1.D. *Let $\varphi \in C_0^\infty(\mathbb{R}^n)$ be supported in $|\xi| < K$. Suppose $\psi \in C_0^\infty(\mathbb{R}^n)$ vanishes for $|\xi| \leq K$. Then again* (D.1.13) *satisfies the estimate* (D.1.3).

PROOF: Under these hypotheses, there exists $\rho \in C_0^\infty(\mathbb{R}^n)$, such that $\rho(0) = 0$ and such that (D.1.14) holds, so again (D.1.13) actually has the form (D.1.3). This time all the hypotheses of Proposition D.1.A are satisfied, so the desired estimate holds.

Further arguments leading to the full proof of Theorem 3.5.A can be found in Chapter 6 of [CM].

Index of notation

We use pseudodifferential operators, associated to symbols via (1.1.12), for symbols $p(x, \xi)$ belonging to a number of symbol spaces. We list these symbol spaces here and record where their definitions can be found. We also use a number of linear spaces of functions and distributions, similarly listed below.

We mention another notational usage. We use $\partial^m u$ to stand for the collection of $\partial^\alpha u = \partial^\alpha u / \partial x^\alpha$, for $|\alpha| = m$, and we use $D^m u$ to stand for the collection $\partial^\alpha u$ for $|\alpha| \le m$. Another common notation for the latter collection is $J^m u$, called the m-jet of u, but I could not bring myself to join the jet set.

symbol	where defined
$S^m_{\rho,\delta}$	(0.1.4)
$XS^m_{1,0}$	(1.1.2)
XS^m_{cl}	(1.1.11)
$C^s_* S^m_{1,\delta}$	(1.3.16)-(1.3.17)
$C^s S^m_{1,\delta}$	(1.3.18)
$\mathcal{A}^r S^m_{1,\delta}$	(3.1.28)
$\mathcal{A}^r_0 S^m_{1,\delta}$	(3.1.31)
${}^r S^m_{1,\delta}$	(3.1.32)
$\tilde{S}^m_{1,1}$	(3.3.34)
$\mathcal{B}^r S^m_{1,1}$	(3.4.1)
Σ^m_r	(3.4.32)

space	where defined
C^s	(1.1.6), (A.1.1)
C^s_*	(1.1.10), (A.1.2)
$H^{s,p}$	(1.1.7), (A.1.7)
$H^{s,p}_{mcl}$	(3.1.51)
BMO	(3.6.30), (A.3.2)
$C^0_\#$	(4.1.18), (B.2.1)
$C^r_\#$	(4.1.21)
$B^0_{\infty,1}$	(4.1.23)
M^p	(A.2.2)
M^p_q	(A.2.10)
C_φ	(9.1.4)
C^m_φ	(9.1.6)

We also include references to some miscellaneous notation of frequent use in the paper.

notation	where defined
$\langle\xi\rangle$	(0.1.4)
$p^{\#}(x,\xi)$	(1.3.2)
$p^{b}(x,\xi)$	(1.3.5)
J_ε	(1.3.3), (5.1.4)
$M(u;x,D)$	(3.1.15), (3.3.6)
$\pi(a,f)$	(3.2.0), (3.2.15)
$T_a f$	(3.2.4)
$R(f,g)$	(3.5.3)
$\tau(a,f)$	(3.5.15)
$T_g^R f$	(3.6.16)
Λ^s	(3.6.19)
$\{c,a\}(x,\xi)$	(6.1.21)
H_a	(6.1.21)

References

[ADN] S.Agmon, A.Douglis, and L.Nirenberg, Estimates near the boundary for solutions of elliptic differential equations satisfying general boundary conditions, CPAM 12(1959), 623-727.

[AC] M.Anderson and J.Cheeger, C^{α}-compactness for manifolds with Ricci curvature and injectivity radius bounded below. Preprint, 1990.

[BKM] T.Beale, T.Kato, and A.Majda, Remarks on the breakdown of smooth solutions for the 3-d Euler equations, Commun. Math. Phys. 94(1984), 61-66.

[Be] M.Beals, Propagation of smoothness for nonlinear second order strictly hyperbolic differential equations, AMS Proc. Symp. Pure Math. 43(1985).

[Be2] M.Beals, Propagation and Interaction of Singularities in Nonlinear Hyperbolic Problems, Birkhauser, 1989.

[BR] M.Beals and M.Reed, Microlocal regularity theorems for nonsmooth pseudodifferential operators and applications to nonlinear problems, TAMS 285 (1984), 159-184.

[Bo] J.Bony, Calcul symbolique et propagation des singularities pour les equations aux derivees nonlineaires, Ann. Sci. Ecole Norm. Sup. 14(1981), 209-246.

[B2] J.Bony, Second microlocalization and propagation of singularities for semi-linear hyperbolic equations, Taniguchi Symp. Katata 1984, pp. 11-49.

[BG] G.Bourdaud, Une algebre maximale d'operateurs pseudodifferentiels, Comm. PDE 13(1988), 1059-1083.

[BrG] H.Brezis and T.Gallouet, Nonlinear Schrödinger evolution equations, J. Nonlin. Anal. 4(1980), 677-681.

[BrW] H.Brezis and S.Wainger, A note on limiting cases of Sobolev imbeddings, Comm. PDE 5(1980), 773-789.

[Br] F.Browder, A priori estimates for elliptic and parabolic equations, AMS Proc. Symp. Pure Math. IV(1961), 73-81.

[CNS] L.Caffarelli, L.Nirenberg, and J.Spruck, The Dirichlet problem for nonlinear second-order elliptic equations I, Monge-Ampere equations, CPAM 37(1984), 369-402.

[Ca1] A.P.Calderon, Commutators of singular integral operators, Proc.NAS,USA 53(1965), 1092-1099.

[Ca2] A.P.Calderon, Singular integrals, Bull. AMS 72(1966), 427-465.

[Ca3] A.P.Calderon, Algebras of singular integral operators, AMS Proc. Symp. Pure Math. X(1967), 18-55.

[Che] J.Chemin, Calcul paradifferentiel precise et applications a des equations aux derivees partielles non semi-lineaires, Duke Math. J. 56(1988), 431-469.

[Che2] J.Chemin, Remarques sur l'existence globale pour le systeme de Navier-Stokes incompressible. Preprint, 1990.

[Ch] M.Christ, Singular integral operators, CBMS Reg. Conf. 1989.

[CK] D.Christodoulu and S.Klainerman, On the local stability of Minkowski spacetime. Preprint, 1990.

[CM] R.Coifman and Y.Meyer, Au dela des operateurs pseudo-differentiels. Asterisque, no.57, Soc. Math de France, 1978.

[CM2] R.Coifman and Y.Meyer, Commutateurs d'integrales singulieres et operateurs multilineaires, Ann. Inst. Fourier Grenoble 28(1978), 177-202.

[DW] F.David and M.Williams, Singularities of solutions to semilinear boundary value problems, Amer. J. Math. 109(1987), 1087-1109.

[DJ] G.David and J.Journe, A boundedness criterion for generalized Calderon-Zygmund operators, Ann. Math. 129(1984), 371-397.

[DeT] D.DeTurk, The Cauchy problem for Lorentz metrics with prescribed Ricci curvature, Comp. Math. 48(1983), 327-349.

[DK] D.DeTurk and J.Kazdan, Some regularity theorems in Riemannian geometry, Ann. Sci. Ecole Norm. Sup. 14(1981), 249-260.

[ES] J.Eells and J.Sampson, Harmonic mappings of Riemannian manifolds, Amer. J. Math. 86(1964), 109-160.

[Ev] L.Evans, Classical solutions of fully nonlinear, convex, second-order elliptic equations, CPAM 35(1982), 333-363.

[FP] C.Fefferman and D.Phong, On positivity of pseudodifferential operators, Proc.NAS,USA 75(1978), 4673-4674.

[FS] C.Fefferman and E.Stein, H^p functions of several variables, Acta Math. 129(1972), 137-193.

[FM] A.Fischer and J.Marsden, General relativity, partial differential equations, and dynamical systems, AMS Proc. Symp. Pure Math. 23(1973), 309-327.

[FL] K.Friedrichs and P.Lax, On symmetrizable differential operators, Proc. Symp. Pure Math. X, AMS, 1967, pp. 128-137.

[GR] P.Gerard and J.Rauch, Propagation de la regularite locale de solutions d'equations hyperboliques non lineaires, Ann. Inst. Fourier Grenoble 37(1987), 65-84.

[GT] D.Gilbarg and N.Trudinger, Elliptic Partial Differential Equations of Second Order, 2nd Ed., Springer-Verlag, NY, 1983.

[H1] L.Hörmander, The Analysis of Linear Partial Differential Operators, vols. 3 and 4, Springer-Verlag, 1985.

[H2] L.Hörmander, Pseudo-differential operators of type 1,1. Comm. PDE 13(1988), 1085-1111.

[H3] L.Hörmander, The Nash-Moser theorem and paradifferential operators. Preprint.

[H4] L.Hörmander, Non-linear Hyperbolic Differential Equations. Lecture Notes, Lund Univ., 1986-87.

[HKM] T.Hughes, T.Kato, and J.Marsden, Well-posed quasi-linear second order hyperbolic systems with applications to nonlinear elastodynamics and general relativity, Arch. Rat. Mech. Anal. 63(1976), 273-294.

[JN] F.John and L.Nirenberg, On functions of bounded mean oscillation, CPAM 14(1961), 415-426.

[JMR] J.Joly, G.Metivier, and J.Rauch, Resonant one dimensional nonlinear geometric optics, Centre de Recherche en Mathematique de Bordeaux, 1990.

[J] J.Jost, Nonlinear Methods in Riemannian and Kahlerian Geometry, Birkhauser, 1988.

[K] T.Kato, Quasi-linear equations of evolution, with applications to partial differential equations, Springer LNM 448(1974), 25-70.

[KP] T.Kato and G.Ponce, Commutator estimates and the Euler and Navier-Stokes equations, CPAM 41(1988), 891-907.

[KS] D.Kinderlehrer and G.Stampacchia, An Introduction to Variational Inequalities and their Applications, Academic Press, NY, 1980.

[KN] H.Kumano-go and M.Nagase, Pseudo-differential operators with nonregular symbols and applications, Funkcial. Ekvac. 21(1978), 151-192.

[La] B.Lascar, Singularites des solutions d'equations aux derivees partielles nonlineaires, CR Acad. Sci. Paris 287A(1978), 527-529.

[L1] P.Lax, The theory of hyperbolic equations. Stanford Lecture Notes, 1963.

[L2] P.Lax, Hyperbolic systems of conservation laws II, CPAM 10(1957), 537-566.

[Lei] E.Leichtnam, Regularite microlocale pour des problemes de Dirichlet nonlineaires noncharacteristiques d'ordre deux a bord peu regulier, Bull. Soc. Math. France 115(1987), 457-489.

[Mj] A.Majda, Compressible Fluid Flow and Systems of Conservation Laws in Several Space Variables, Appl. Math. Sci. #53, Springer, 1984.

[Mj2] A.Majda, The stability of multi-dimensional shock fronts. Memoirs AMS, no.275, 1983.

[Mj3] A.Majda, The existence of multi-dimensional shock fronts. Memoirs AMS, no.281, 1983.

[Mj4] A.Majda, Nonlinear geometric optics for hyperbolic systems of conservation laws, Univ. of Minn. IMA Preprint, no. 202, 1985.

[Ma] J.Marschall, Pseudo-differential operators with coefficients in Sobolev spaces, TAMS 307(1988), 335-361.

[MeR] R.Melrose and N.Ritter, Interaction of nonlinear progressing waves, Ann. Math. 121(1985), 187-213.

[Met1] G.Metivier, Interaction de deux chocs pour un systeme de deux lois de conservation en dimension deux d'espace, TAMS 296(1986), 431-479.

[Met2] G.Metivier, Stability of multi-dimensional weak shocks, Comm. PDE 15(1990), 983-1028.

[M1] Y.Meyer, Regularite des solutions des equations aux derivees partielles non lineaires, Springer LNM 842(1980), 293-302.

[M2] Y.Meyer, Remarques sur un theoreme de J.M.Bony, Rend. del Circolo mat. di Palermo (suppl. II:1) (1981), 1-20.

[M3] Y.Meyer, Real analysis and operator theory, Proc. Symp. Pure Math. 43(1985), 219-235.

[M4] Y.Meyer, Wavelets and operators, in Analysis at Urbana 1, E. Berkson et al. eds., London Math. Soc. Lecture Notes, no. 137, 1989, pp. 256-365.

[Mor] C.B.Morrey, Multiple integrals in the calculus of variations, Springer-Verlag, 1966.

[Mo] J.Moser, A rapidly convergent iteration method and nonlinear partial differential equations, I, Ann. Scuola Norm. Sup. Pisa 20(1966), 265-315.

[Mo2] J.Moser, A new proof of de Giorgi's theorem concerning the regularity problem for elliptic differential equations, CPAM 13(1960), 457-468.

[Mo3] J.Moser, On Harnack's theorem for elliptic differential equations, CPAM 14(1961), 577-691.

[Na] M.Nagase, The L^p boundedness of pseudodifferential operators with non-regular symbols, Comm. PDE 2(1977), 1045-1061.

[Ni] L.Nirenberg, On elliptic partial differential equations, Ann. Scuola Norm. Sup. Pisa 13(1959), 116-162.

[Ni2] L.Nirenberg, Estimates and existence of solutions of elliptic equations, CPAM 9(1956), 509-530.

[Po] J.Polking, Boundary value problems for parabolic systems of partial differential equations, AMS Proc. Symp. Pure Math. X(1967), 243-274.

[Ra] J.Rauch, Singularities of solutions to semilinear wave equations, J. Math. Pures et Appl. 58(1979), 299-308.

[RR] J.Rauch and M.Reed, Propagation of singularities for semilinear hyperbolic equations in one space variable, Ann. Math. 111(1980), 531-552.

[RR2] J.Rauch and M.Reed, Nonlinear microlocal analysis of semilinear hyperbolic systems in one space dimension, Duke Math. J. 49(1982), 397-475.

[RR3] J.Rauch and M.Reed, Singularities produced by the nonlinear interaction of three progressing waves: examples, Comm. PDE 7(1982), 1117-1133.

[RR4] J.Rauch and M.Reed, Classical conormal solutions to semilinear systems, Comm. PDE 13(1988), 1297-1335.

[ST] M.Sable-Tougeron, Regularite microlocale pour des problemes aux limites non lineaires, Ann. Inst. Fourier 36(1986), 39-82.

[ST2] M.Sable-Tougeron, Propagation des singularities faibles en elastodynamique non lineaire, J. Math. Pures et Appl. 67 (1988), 291-310.

[Se] J.Serrin, The problem of Dirichlet for quasilinear elliptic differential equations with many independent variables, Phil. Trans. Royal Soc. London Ser. A 264(1969), 413-496.

[S1] E.Stein, Singular Integrals and Differentiability Properties of Functions, Princeton Univ. Press, 1970.

[S2] E.Stein, Singular Integrals and Pseudo-differential Operators, Graduate Lecture Notes, Princeton Univ., 1972.

[Str] M.Struwe, Variational Problems, Springer-Verlag, 1990.

[T1] M.Taylor, Pseudodifferential Operators, Springer LNM 416, 1974.

[T2] M.Taylor, Pseudodifferential Operators, Princeton Univ. Press, 1981.

[T3] M.Taylor, Fefferman-Phong inequalities in diffraction theory, Proc. Symp. Pure Math. 43(1985), 261-300.

[Tr] H.Triebel, Theory of Function Spaces, Birkhauser, Boston, 1983.

[Tru] N.Trudinger, Fully nonlinear, uniformly elliptic equations under natural structure conditions, TAMS 278(1983), 751-769.

[W] M.Williams, Spreading of singularities at the boundary in semilinear hyperbolic mixed problems, I, Duke Math. J. 56(1988), 17-40; II, Trans. AMS, to appear.

Progress in Mathematics

Edited by:

J. Oesterlé
Département de Mathématiques
Université de Paris VI
4, Place Jussieu
75230 Paris Cedex 05, France

A. Weinstein
Department of Mathematics
University of California
Berkeley, CA 94720
U.S.A.

Progress in Mathematics is a series of books intended for professional mathematicians and scientists, encompassing all areas of pure mathematics. This distinguished series, which began in 1979, includes authored monographs and edited collections of papers on important research developments as well as expositions of particular subject areas.

We encourage preparation of manuscripts in some form of TeX for delivery in camera-ready copy which leads to rapid publication, or in electronic form for interfacing with laser printers or typesetters.

Proposals should be sent directly to the editors or to: Birkhäuser Boston, 675 Massachusetts Avenue, Cambridge, MA 02139, U. S. A.

A complete list of titles in this series is available from the publisher.

53 LAURENT. Théorie de la Deuxième Microlocalisation dans le Domaine Complexe.
54 VERDIER/LE POTIER. Module des Fibres Stables sur les Courbes Algébriques: Notes de l'Ecole Normale Supérieure, Printemps, 1983
55 EICHLER/ZAGIER. The Theory of Jacobi Forms
56 SHIFFMAN/SOMMESE. Vanishing Theorems on Complex Manifolds
57 RIESEL. Prime Numbers and Computer Methods for Factorization
58 HELFFER/NOURRIGAT. Hypoellipticité Maximale pour des Opérateurs Polynomes de Champs de Vecteurs
59 GOLDSTEIN. Séminaire de Théorie des Nombres, Paris 1983–84
60 PROCESI. Geometry Today: Giornate Di Geometria, Roma. 1984
61 BALLMANN/GROMOV/SCHROEDER. Manifolds of Nonpositive Curvature
62 GUILLOU/MARIN. A la Recherche de la Topologie Perdue
63 GOLDSTEIN. Séminaire de Théorie des Nombres, Paris 1984–85
64 MYUNG. Malcev-Admissible Algebras
65 GRUBB. Functional Calculus of Pseudo-Differential Boundary Problems
66 CASSOU-NOGUES/TAYLOR. Elliptic Functions and Rings and Integers
67 HOWE. Discrete Groups in Geometry and Analysis: Papers in Honor of G.D. Mostow on His Sixtieth Birthday
68 ROBERT. Autour de L'Approximation Semi-Classique
69 FARAUT/HARZALLAH. Deux Cours d'Analyse Harmonique

70 ADOLPHSON/CONREY/GHOSH/YAGER. Analytic Number Theory and Diophantine Problems: Proceedings of a Conference at Oklahoma State University
71 GOLDSTEIN. Séminaire de Théorie des Nombres, Paris 1985–86
72 VAISMAN. Symplectic Geometry and Secondary Characteristic Classes
73 MOLINO. Riemannian Foliations
74 HENKIN/LEITERER. Andreotti-Grauert Theory by Integral Formulas
75 GOLDSTEIN. Séminaire de Théorie des Nombres, Paris 1986–87
76 COSSEC/DOLGACHEV. Enriques Surfaces I
77 REYSSAT. Quelques Aspects des Surfaces de Riemann
78 BORHO /BRYLINSKI/MACPHERSON. Nilpotent Orbits, Primitive Ideals, and Characteristic Classes
79 MCKENZIE/VALERIOTE. The Structure of Decidable Locally Finite Varieties
80 KRAFT/PETRIE/SCHWARZ. Topological Methods in Algebraic Transformation Groups
81 GOLDSTEIN. Séminaire de Théorie des Nombres, Paris 1987–88
82 DUFLO/PEDERSEN/VERGNE. The Orbit Method in Representation Theory: Proceedings of a Conference held in Copenhagen, August to September 1988
83 GHYS/DE LA HARPE. Sur les Groupes Hyperboliques d'après Mikhael Gromov
84 ARAKI/KADISON. Mappings of Operator Algebras: Proceedings of the Japan-U.S. Joint Seminar, University of Pennsylvania, Philadelphia, Pennsylvania, 1988
85 BERNDT/DIAMOND/HALBERSTAM/HILDEBRAND. Analytic Number Theory: Proceedings of a Conference in Honor of Paul T. Bateman
86 CARTIER/ILLUSIE/KATZ/LAUMON/MANIN/RIBET. The Grothendieck Festschrift: A Collection of Articles Written in Honor of the 60th Birthday of Alexander Grothendieck. Vol. I
87 CARTIER/ILLUSIE/KATZ/LAUMON/MANIN/RIBET. The Grothendieck Festschrift: A Collection of Articles Written in Honor of the 60th Birthday of Alexander Grothendieck. Volume II
88 CARTIER/ILLUSIE/KATZ/LAUMON/MANIN/RIBET. The Grothendieck Festschrift: A Collection of Articles Written in Honor of the 60th Birthday of Alexander Grothendieck. Volume III
89 VAN DER GEER/OORT / STEENBRINK. Arithmetic Algebraic Geometry
90 SRINIVAS. Algebraic K-Theory
91 GOLDSTEIN. Séminaire de Théorie des Nombres, Paris 1988–89
92 CONNES/DUFLO/JOSEPH/RENTSCHLER. Operator Algebras, Unitary Representations, Envelopping Algebras, and Invariant Theory. A Collection of Articles in Honor of the 65th Birthday of Jacques Dixmier
93 AUDIN. The Topology of Torus Actions on Symplectic Manifolds
94 MORA/TRAVERSO (eds.) Effective Methods in Algebraic Geometry
95 MICHLER/RINGEL (eds.) Represenation Theory of Finite Groups and Finite Dimensional Algebras
96 MALGRANGE. Equations Différentielles à Coefficients Polynomiaux
97 MUMFORD/NORI/NORMAN. Tata Lectures on Theta III
98 GODBILLON. Feuilletages, Etudes géométriques
99 DONATO /DUVAL/ELHADAD/TUYNMAN. Géométrie Symplectique et Physique Mathématique. A Collection of Articles in Honor of the 65th Birthday of J.-M. Souriau
100 TAYLOR. Pseudodifferential Operators and Nonlinear PDE